Economic Geography

During the height of the 'quantitative revolution' of the 1960s, Economic Geography was a tightly focused and specialized field of research. Now, it sprawls across several disciplines to embrace multiple theoretical, philosophical and empirical approaches. This volume moves economic geography through a series of theoretical and methodological approaches, looking both towards the future and to the discipline's engagement with public policy.

Economic Geography covers contributions by selected economic geographers whose purpose is to help explain the interconnection among all forces that trigger societal change, namely the ever-changing capitalist system. The contributors record changing foci and methodologies from the 1960–1980 period of quantitative economic geography, the 1980s interest in understanding how regimes of accumulation in a capitalist world construct spaces of uneven development, and how the 1990s literature was enriched by differing viewpoints and methodologies which were designed to understand the local effects of the global space economy. In the new century, the overwhelming response has been that of bridging gaps across 'voices within the sub-discipline of Economic Geography' in order to maximize our understanding of processes that shape our social, political and economic existence. Contributors also highlight what they see as the challenges for understanding contemporary issues, thus putting down markers for younger researchers to take the lead on.

Through a collection of 20 chapters on theoretical constructs and methodologies, debates and discourses, as well as links to policymaking and policy evaluation, this volume provides a succinct view of concepts and their historical trajectories in Economic Geography. Readers are exposed to the breadth of the discipline and engaged in current debates and understandings of the critical components of research in economic geography, theoretical, empirical and applied.

Sharmistha Bagchi-Sen is a Professor in the Department of Geography, University at Buffalo-State University of New York.

Helen Lawton Smith is Reader in Management, School of Management and Organisational Psychology, Birkbeck, London University and a Distinguished Research Associate at the School of Geography, Oxford University.

Economic Geography

Past, present and future

Edited by
Sharmistha Bagchi-Sen and Helen Lawton Smith

First published 2006 by Routledge

Published 2017 by Routledge
2 Park Square, Milton Park, Abingdon, Oxon OX14 4RN
711 Third Avenue, New York, NY 10017, USA

First issued in paperback 2014

Routledge is an imprint of the Taylor & Francis Group, an informa business

Typeset in Galliard by Keyword Group Ltd

British Library Cataloguing in Publication Data
A catalogue record for this book is available from the British Library

Library of Congress Cataloging in Publication Data
A catalog record for this book has been requested.

ISBN 13: 978-1-138-86715-4 (pbk)
ISBN 13: 978-0-415-36784-4 (hbk)

Routledge Studies in Economic Geography

The Routledge Studies in Economic Geography series provides a broadly based platform for innovative scholarship of the highest quality in economic geography. Rather than emphasizing any particular sub-field of economic geography, we seek to publish work across the breadth of the field and from a variety of theoretical and methodological perspectives.

Published:
Economic Geography: Past, present and future
Edited by Sharmistha Bagchi-Sen and Helen Lawton Smith

Forthcoming:
Remaking Regional Economies: Firm strategies, labor markets and new uneven development
Susan Christopherson and Jennifer Clark

The New Economy of the Inner City: Regeneration and dislocation in the twenty first century metropolis
Thomas Hutton

Contents

International Advisory Board

Series Preface

Over the past half century, the field of economic geography has been marked by periods of particular dynamism and innovation. From the quantitative revolution of the 1960s to the emergence of a new industrial geography during the 1980s, a combination of theoretical innovation and rapidly changing economic circumstance have made for an intellectually dynamic field of enquiry. The past decade has been no less significant in terms of theoretical and empirical advance. Economic geography today is a vibrant and growing field of study. New lines of research are emerging that build upon a broadened concept of the economic, upon analysis of economic development and global economic change, and upon renewed interest in issues of policy, institutions and governance. Longstanding research interests in industrial and technological change are being vigorously pursued in the context of new theories of learning and innovation. Economic geography today is methodologically diverse, engaged with issues of compelling social concern, and alive with interesting and provocative scholarship.

We are delighted in this context to support the launch of *Routledge Studies in Economic Geography*. The intent of this new book series is to provide a broadly based platform for innovative scholarship of the highest quality in economic geography. Rather than emphasizing any particular sub-field of economic geography, we seek to publish work across the breadth of the field and from a variety of theoretical and methodological perspectives. In launching the book series, we also seek to support and promote a move toward a broader, more integrated economic geography. Economic geography now reaches into domains of culture, gender, governance, and nature-society relations that heretofore typically have been treated more or less as separate domains of enquiry. Arguably, some of the most exciting work within economic geography today lies at these interfaces of economic change, whether this is in terms of cultural construction of economies, or the relationship between industrial development, resources and the environment.

Contemporary processes of global economic change are also stimulating new research agendas in economic geography. Exciting new research is emerging around the scalar dynamics and relational geographies of global economic change, including work on such topics as global organizations and global development policies, deregulation of markets and investment regimes and attendant consequences for sustainable livelihoods around the world, and the local and regional development dynamics accompanying intensified flows of capital, technology and information on a global scale. One consequence of these processes of economic change is that the predominant focus of economic geography on OECD economies is now giving way to a more 'global' economic geography in which existing boundaries with 'development geography' and 'area studies' are giving way. Indeed, it makes little sense to talk of an economic geography absent analysis of developing economies and economies in transition. By the same token, research into the economic geographies of these regions is becoming a source for further theoretical innovation within the field.

Many positive developments are underway that help feed economic geography as a vibrant field of enquiry. We note with pleasure the emergence of new journals and the widespread support for a summer institute that exposes graduate students and early career faculty to the very latest theoretical and methodological developments within the field. We also welcome the engagement across academic disciplines and among scholarly networks that marks much cutting-edge research in economic geography. The field is also supported by the availability of publishing platforms that actively promote the bringing to fruition of sustained periods of scholarship in the form of book manuscripts. In an era of shortened cycles of research and publication, there remains an important role for book manuscripts that bring together the cumulative results of sustained programmes of research, theoretical innovation and empirical investigation. *Routledge Studies in Economic Geography* seeks to provide such a publishing platform for innovative scholarship of the highest quality across the breadth of the field of economic geography. We hope that the volumes in this series will inspire further theoretical and methodological innovation, as well as new insights into economic welfare, livelihoods and the dynamics of economic change locally and around the world.

David P. Angel
Clark University, USA
Amy K. Glasmeier
Pennsylvania State University, USA
Adam Tickell
University of Bristol, UK
June 2006

Foreword

'The real duty of the ... [economic geographer] ... is not ... [just] ... to explain our sorry reality, but to improve it' (Lösch 1954: 4). I use this paraphrased quotation from a no longer fashionable 'location theorist' for four reasons. First, it was primarily because of Lösch's book, *The Economics of Location* (introduced to me as an undergraduate in the early 1960s by David M. Smith), that I became an economic geographer. What attracted me to his complex book – and it seemed *very* complex indeed for somebody fed on the indigestible descriptive texts of traditional economic geography that constituted the conventional diet of economic geography courses – was that its heavy *theoretical* orientation was combined with the deep sense of *social concern*. The purpose of any academic discipline worth its salt must be to try to improve the world in which we live: to engage with real problems, to 'get our hands dirty' in Amy Glasmeier's words. Of course, the kind of theoretical approach espoused by Lösch soon became outmoded and displaced, as several of the contributors to this volume explain, by a succession of alternative approaches, each being heralded as the new orthodoxy by its adherents. But the normative question of 'what ought to be' – as well as 'what is' – should still be at the centre of our concerns as economic geographers.

The second reason I refer to Lösch here is that his approach first made me appreciate the importance of theoretically-informed empirical research. His book was surprisingly rich in a wide range of empirical materials at different spatial scales. Of course, I no longer espouse the particular theoretical framework of the economic location theorists but that is not the point. We need to develop theories to help us make sense of the world and those theories need to be firmly grounded. Today economic geographers engage in far more diverse theoretical explanations than in the 1960s and I am sure this is a good thing. The world is far too complex to be captured by a single over-arching theoretical framework. As Ray Hudson argues, 'the economic geographies of the late modern capitalist world are too complex and nuanced to be explicable in terms of one all-encompassing theoretical position'.

The diversity of contemporary economic geography is – or should be – a strength, not a weakness (see Ann Markusen's and Eric Sheppard's chapters). But we do need to build theories; mere description is not enough. Such theories need to be able to incorporate the complex, and highly unequal, power-laden interactions between the multifarious sets of actors, agents and institutions that constitute economies and the ways in which these stretch across a continuum of geographical scales and, at the same time, inter-penetrate 'territories'. In my own view, such theorizations should include a strong focus on the relationalities of situated networks. As Katherine Mitchell points out, 'thinking in terms of networks forces us to theorize socioeconomic processes as intertwined and mutually constitutive'. But we must always remember that such networks are not independent of the macro-structural frameworks within which they are embedded and with which they continuously engage in dialectical interaction. We are all, to a large degree, involved in 'political economy'.

On the other hand, we also need careful, robust, well-designed empirical research. We need, in other words, to focus not only on processes but also on outcomes. And we need to do so using techniques appropriate to the task, whether these are qualitative or quantitative or a mix of the two. But both our theories and our empirical work need to avoid the ethnocentrism that is characteristic of most economic–geographical research, embedded as it tends to be in the western (especially Anglo-American) industrialized countries (see Henry Yeung's chapter).

We also need to broaden our investigative horizons in terms of the phenomena we study. Some kinds of economic activity attract a disproportionate amount of our attention; others are virtually ignored or, at best, under-researched. There are significant 'silences'. Much of the work continues to be heavily productionist, with very little real integration of processes of consumption in our analyses. Within manufacturing, there continues to be a narrow focus on a few specific sectors and a neglect of others. A similar criticism can be made of work on services. We know a lot about financial services, for example, but logistical services are virtually ignored. Agriculture continues to be given inadequate attention within economic geography, despite the fact that this sector employs vast numbers of people in developing countries and is one of the most sensitive issues in current globalization debates. In a related vein, few economic geographers today research natural resources and far too few economic geographers have developed a serious engagement with environmental issues. Of course, there are honourable exceptions in all of these cases – many of them represented in this book – but I think the overall criticism is justified.

Third, my engagement with Lösch in the 1960s first took me beyond the boundaries of geography and into the realms of another social science: economics. There is much lively debate today about whether, and how, economic geographers should engage with economists. I go along with the view of several contributors to this book that economic geography must engage with economics, but that is 'never sufficient' (Richard Walker). We need to engage in productive dialogue with all the relevant disciplines, but to do so in ways that build upon economic geography's own strengths. Eric Sheppard writes of creating 'trading zones' between different disciplinary approaches and I very much agree with that. But an academic discipline's success in trading depends on its own internal strength and distinctive identity. Without a strong disciplinary core, there would be little to trade with others. There is always the danger of economic geography (indeed of geography as a whole) being swallowed up. One of economic geography's undoubted strengths, as Ann Markusen points out, is its synthesizing abilities across the social and natural sciences.

But while we undoubtedly need to engage productively with non-geographers we also need to engage more with other sub-fields within geography itself. Most obviously this is true of the long-standing human-physical divide, one that is so self-evidently debilitating in the context of global environmental problems. But it is also true, for example, of the lack of real connections between economic geographers and development geographers. In such an uneven world – and of the need to improve our 'sorry reality' – this is not just stupid, it is bordering on the criminal.

My fourth reason for recalling Lösch is that it reminds us of the importance of having a real sense and understanding of the history of economic geography as a distinctive sub-field within geography and within the social sciences in general. Economic geography, like geography as a whole, has a history of ignoring its history; of not just discarding theoretical frameworks or methodologies but of writing them completely out of the script. As both Susan Hanson and Ann Markusen point out, this is very short-sighted. We need to know where we come from; we need to understand why approaches have changed. As Susan Hanson argues: 'we draw upon the past to envision the future . . . The ease with which authors fail to link their own work to earlier work . . . simply does not make sense to me because it means that much of value is needlessly discredited, submerged and lost . . . a look at the history of this field provokes a call for greater ecumenism, for more willingness to see the connections across the decades, and for the enduring tolerance that making those connections should foster.'

The chapters in this book exemplify each of these concerns. They provide valuable and stimulating perspectives on how and why economic geographers do what they do in ways that demonstrate the values that economic geographers can bring to explaining and helping to improve the lives of people and communities wherever they are. It is a challenging agenda but one that must be grasped. As Lösch said, that is our 'real duty'.

Peter Dicken
University of Manchester, UK

Acknowledgements

We first acknowledge the help and patience of our colleagues, Zoe Kruze and Andrew Mold at Routledge in seeing this project through. We also thank the anonymous referees who helped us define more closely the nature of the book. We are very grateful to Peter Dicken for kindly writing the preface. It was *Location in Space*, written with Peter Lloyd that provided the inspiration for the five sessions at the Centennial AAG. Very special thanks are owed to Professor Gordon Clark for his kind help and support in organizing the sessions and for his subsequent advice on the book. We are also very grateful to Doug Watts who was involved at the early stages of planning the sessions. We also thank the contributors for responding to our requests with such good humour. Finally we would like to thank Annitra Jongsthapongpanth for assistance with the manuscript.

Contributors

David P. Angel is a Professor in the School of Geography at Clark University where he also holds the Leo L. and Joan Kraft Laskoff Professorship in Economics, Technology and the Environment. An economic geographer by training, Professor Angel's research focuses on issues of economic change and the environment. Recent books include *Asia's Clean Revolution: Industry Growth and the Environment* (Greenleaf Publishing, 2000) and *Industrial Transformation in the Developing World* (Oxford University Press, 2005), both co-authored with economist Michael Rock. Professor Angel's current position is Provost and Vice President for Academic Affairs at Clark University.

Bjørn T. Asheim is Professor of Economic Geography at Lund University, Sweden, and Professor II at the Centre of Technology, Innovation, and Culture, University of Oslo, Norway. He is co-founder and deputy director of the Centre of Excellence in Innovation System Research (CIRCLE) at Lund University. He has recently co-edited a book on *Regional Innovation Policy for Small-Medium Enterprises* (Edward Elgar, 2003), has co-edited *Clusters and Regional Development* (Routledge, 2006) with Phil Cooke and Ron Martin. He is the author of numerous articles and book chapters on industrial districts, regional innovation systems and learning regions. He is Editor of *Economic Geography* and member of the editorial board of *European Planning Studies*, and the *Journal of Economic Geography*. His current research is on the geography of the creative class and regional innovation systems.

Sharmistha Bagchi-Sen is a Professor of Geography at the State University of New York at Buffalo. She received her PhD from the University of Georgia in 1989. Her research has focused on multinationals and foreign direct investment, export market development by small firms, producer services, innovation and collaboration in high technology industry, the biotechnology industry in the United States, and labour market issues in producer services and information technology. She currently serves as the editor of *The Professional Geographer* and is a member of the editorial board of the *Annals of the Association of American Geographers*.

William B. Beyers is a Professor of Geography at the University of Washington. He earned his BA at the University of Washington in 1962, and his PhD (1967) is also from the University of Washington. He has been on the faculty since 1967. His research has focused on various topics in the field of economic geography, including regional economic models, national trends in the distribution of economic activity in the United States, the development of producer services and cultural industries. He has also undertaken numerous economic impact studies, on topics ranging from major league sports to arts and cultural organizations.

Gordon L. Clark is the Halford Mackinder Professor of Geography and Head of the Oxford University Centre for the Environment, and Senior Research Associate in the Labor and Worklife Program at Harvard Law School, Harvard University. An economic geographer with continuing interest in the provision of urban infrastructure, his current research is on institutional governance, decision-making, and pension policy. Recent publications include *The Geography of Finance, Pension Security in the 21st Century* (Oxford University Press, 2004), and *European Pensions and Global Finance* (Oxford University Press, 2003).

Peter W. Daniels is Professor of Geography and Co-Director, Services and Enterprise Research Unit, University of Birmingham (UK). He has undertaken research on the service economy, especially producer services as key agents in metropolitan and regional restructuring at the national and international scale. His publications include *Service Industries: A Geographical Appraisal* (1985), *Services and Metropolitan Development* (1991), *Service Industries in the World Economy* (1993), *Services in the Global Economy, Vols I and II* (with J. R. Bryson, 1998), *Service Worlds: People, Organizations, Technologies* (with J. R. Bryson and B. Warf, 2004); *Service Industries and Asia-Pacific Cities: New Development Trajectories* (with K. C. Ho and T. A. Hutton, 2005); *The Service Industries Handbook* (with J. R. Bryson, in press, 2006).

Peter Dicken is Emeritus Professor of Economic Geography in the School of Environment and Development at the University of Manchester, UK. He has held visiting academic appointments at universities and research institutes in Australia, Canada, China, Hong Kong, Mexico, Singapore, Sweden and the US and lectured in many other countries throughout Europe and Asia. He is an Academician of the Social Sciences, a recipient of the Victoria Medal of the Royal Geographical Society (with the Institute of British Geographers) and of an Honorary Doctorate of the University of Uppsala, Sweden.

Rafiq Dossani is a senior research scholar at Shorenstein APARC, responsible for developing and directing the South Asia Initiative. His research interests include South Asian security, and financial, technology, and energy-sector reform in India. He is currently undertaking projects on political reform, business process outsourcing, innovation and entrepreneurship in information technology in India, and security in the Indian subcontinent. His most recent books are *Prospects* *for Peace in South Asia* (co-edited with Henry Rowen), published in 2005 by Stanford University Press, and *Telecommunications Reform in India*, published in 2002 by Greenwood Press.

Amy K. Glasmeier is the E. Willard Miller Professor of Economic Geography and the John Whisman Scholar of the Appalachian Regional Commission. She is a Professor of Geography and Regional Planning at The Pennsylvania State University. Published in fall 2005 by Routledge, *An Atlas of Poverty in America: One Nation, Pulling Apart 1960–2003*, examines the experience of people and places in poverty since the 1960s, looks across the last four decades at poverty in America and recounts the history of poverty policy since the 1940s. Glasmeier has worked all over the world, including Japan, Hong Kong, Latin America and Europe. She has worked with the OECD, ERVET Emilia Romagna Regional Planning Agency. She is currently engaged in a retrospective examination of poverty and poverty policy history in the US. The work is leading to new perspectives on the nature and extent of persistent poverty in America and is exploring the theoretical and ideological basis for federal poverty policy since the 1960s.

Anne Green has a first degree in Geography from University College London. After postgraduate study she held research posts at the Centre for Urban and Regional Development Studies, University of Newcastle upon Tyne and at the University of Cardiff. She is currently a Principal Research Fellow at the Institute for Employment Research – a multi-disciplinary research centre at the University of Warwick. Her research is primarily concerned with spatial dimensions of economic, social and demographic change; aspects of local and regional labour markets; migration and commuting; and urban, rural and regional development. She has extensive experience of undertaking policy-relevant research for the UK Government.

Susan Hanson is the Jan and Larry Landry University Professor and Professor of Geography at Clark University, where she has taught since 1981. Her teaching and research interests lie at the intersection of urban, economic, and social geography and in feminist geography. With colleagues, she has investigated the activity patterns of urban residents and the role of gender in shaping urban labour markets; she is currently completing a project on gender, geography, and entrepreneurship.

Ray Hudson is Professor of Geography and Director of the inter-disciplinary Wolfson Research Institute at Durham University. He is a member of the Academy of Learned Societies in the Social Sciences and a past-Vice President and Chair of the Research Division of the Royal Geographical Society. His main research interests focus on geographies of economies and territorial development strategies. Current research includes projects on the labour market experiences of former coal miners and steel workers in the UK and Canada, ethnographies of social economies and the production and management of wastes. He holds the degrees of BA, PhD and DSc from Bristol University and an honorary DSc from Roskilde University. In 2005 he was awarded the Victoria Medal by the Royal Geographical Society.

Martin Kenney is a professor at the University of California, Davis. His research interests include the changing economic geography of global capitalism and he has done research in North America, China, and India. He is currently working on the history and globalization of the venture capital industry and the phenomenon of services offshoring. With co-authors he has published in the *American Sociological Review*, *Economic Geography*, *Industrial and Corporate Change*, *Regional Studies*, and *Research Policy*. His most recent edited book *Locating Global Advantage* is published by Stanford University Press where he edits a book series on technology, innovation and the global economy. He has been a visiting professor at Cambridge University, Copenhagen Business School, Hitotsubashi University, Kobe University, Osaka City University, and Tokyo University.

Helen Lawton Smith is Reader in Management at the School of Management and Organisational Psychology, Birkbeck, University of London. She is a Distinguished Research Associate at the Department of Geography, Oxford University. She is the founder and Director of Research of the Oxfordshire Economic Observatory, Oxford University. Her research interests are based on geographies of innovation and include entrepreneurship, technology transfer and scientific labour markets. Recent publications include *Technology Transfer and Industrial Change in Europe* (Palgrave, 2000) and *Universities, Innovation and the Economy* (Routledge, 2006).

John Lovering is Professor of Urban Development and Governance at the School of City and Regional Planning, Cardiff University. After graduating in Economics, and a brief musical career he worked at the School for Advanced Urban Studies of Bristol University, Liverpool University Geography Dept, and Hull University School of Geography, before moving to Cardiff where he set up the Geography and Planning Degree. His research interests include the philosophy of social science, globalization and defence industrial restructuring, urban and regional development and labour market policy.

Linda McDowell is Professor of Human Geography at Oxford University. She has also worked at the Open University, Cambridge, the LSE and University College London. Her main research interest is the interconnections between economic change, new forms of work in the labour market and in the home and the transformation of gender relations in contemporary Britain. She has published widely in this area including several books: *Capital Culture: Gender at Work in the City* (Blackwell, 1997), *Gender, Identity and Place* (Polity, 1999), *Redundant Masculinities?* (Blackwell, 2003), *Hard Labour* (UCL Press, 2005), and numerous articles.

Edward J. Malecki is Professor of Geography at The Ohio State University. He has been a Visiting Researcher in the Centre for Urban and Regional Development Studies (CURDS) at the University of Newcastle upon Tyne, England, and a Fulbright Fellow at the University of Economics and Business Administration in Vienna, Austria. He is author of over 100 published papers, and of *Technology and Economic Development* (Addison-Wesley Longman, 1997), is co-editor of *Making Connections: Technological Learning and Regional Economic Change* (Ashgate, 1999). He is Associate Editor of the journal *Entrepreneurship and Regional Development.*

Ann Markusen is Professor and Director of the Project on Regional and Industrial Economics at the Humphrey Institute of Public Affairs, University of Minnesota. Her books include *From Defense to Development; Second Tier Cities; Arming the Future; Trading Industries, Trading Regions; Regions: the Economics and Politics of Territory*; and *Profit Cycles, Oligopoly and Regional Development.* She has served as President of the North American Regional Science Association, Senior Fellow at the Council on Foreign Relations, and Chair of the Committee on Science, Engineering and Public Policy of the American Association for the Advancement of Science.

Ron Martin is Professor of Economic Geography in the University of Cambridge, UK. His research interests include the geographies of economic growth, finance, and labour; and the application of different schools of economic thought to economic geography. He has published 25 books and more than 150 papers on these and related themes. His work has a strong commitment to public relevance and policy. He has edited *Transactions of the Institute of British Geographers*, and *Regional Studies*; and is presently an editor on the *Cambridge Journal of Economics* and on the *Journal of Economic Geography.* Ron is an Academician of the British Academy of Social Sciences, and a Fellow of the British Academy.

Allen J. Scott is Professor of Urban Planning and holds a joint appointment in the Department of Policy Studies and the Department of Geography at UCLA. He is the Director of the Center for Globalization and Policy Research in the School of Public Policy and Social Research. Dr Scott's recent research and writing have been focused on issues of industrialization, urban and regional growth, and globalization. Dr Scott was awarded a Guggenheim Fellowship in 1986–7, and was given the Honors Award of the Association of American Geographers in 1987. In 1999 he was elected a corresponding fellow of the British Academy.

Eric Sheppard is Professor of Geography, with adjunct appointments in the Interdisciplinary Center for Global Change and American Studies, at the University of Minnesota. He has co-authored *The Capitalist Space Economy* (with T. J. Barnes, 1990), *A World of Difference* (with P. W. Porter, 1998), co-edited *A Companion to Economic Geography* (with T. J. Barnes, 2000) and *Scale and*

Geographic Inquiry (with R. B. McMaster, 2004), and published 90 refereed articles and book chapters. Current research interests include the spatiality of capitalism and globalization, international trade, environmental justice, critical GIS, and contestations of neoliberal urbanization.

Henry Wai-Chung Yeung (PhD, University of Manchester) is Professor of Economic Geography at the Department of Geography, National University of Singapore. He is the sole author of three monographs and editor/co-editor of another four books. He has over 65 research papers published in internationally refereed journals and 20 chapters in books. He co-edits three journals (*Environment and Planning A*, *Economic Geography*, and *Review of International Political Economy*) and is Asia-Pacific Editor of *Global Networks*, and Business Manager of *Singapore Journal of Tropical Geography*. He sits on the editorial boards of another seven international journals, including *Asia Pacific Journal of Management*, *European Urban and Regional Studies*, and *Journal of Economic Geography*.

Richard Walker is Professor and past Chair of Geography at the University of California, Berkeley, where he has taught since 1975. He has written scores of articles on diverse topics in economic, urban, and environmental geography. He is author, with Michael Storper, of *The Capitalist Imperative: Territory, Technology and Industrial Growth* (Blackwell, 1989) and, with Andrew Sayer, of *The New Social Economy: Reworking the Division of Labor* (Blackwell, 1992). Recently, his focus has been on the peculiarities of California – one of the most important economic, political and cultural hearths of capitalism. His latest book is *The Conquest of Bread: 150 Years of California Agribusiness* (New Press, 2004) and another is forthcoming, *The Country in the City: The Greening of the San Francisco Bay Area* (University of Washington Press, 2006). A third volume, *Bay City: The Urbanization of the San Francisco Bay Area*, is nearly complete.

H. Doug Watts is a Reader in Geography at the University of Sheffield. He has written extensively on the behaviour of multi-locational firms and has had a special interest in plant closures. This research has been supplemented by the analysis of the behaviour of small firms in clusters of traditional industries. He is the author of *The Large Industrial Enterprise* (Croom Helm, 1979), *The Branch Plant Economy* (Longman, 1981) and *Industrial Geography* (Longman, 1987). He has published in leading journals including *Regional Studies*, *Urban Studies*, *Environment and Planning A* and *Progress in Human Geography*. His main teaching interests have been the economic geography of advanced economies and, especially, industrial geography.

Introduction

The past, present and future of economic geography

Sharmistha Bagchi-Sen and Helen Lawton Smith

This book draws its inspiration from five sessions on economic geography organized by the editors at the Centennial Meeting of the Association of American Geographers (AAG) in Philadelphia in 2004. In the sessions, titled 'Economic Geography: Then, Now and the Future', through a discussion of their own research histories, the panelists were asked to reflect upon the progress in theory and practice of economic geography. The panels were motivated by the recent discourses in economic geography – this reveals that geography students need exposure to various perspectives in order to understand the interconnection among all forces that trigger societal change, namely the ever-changing capitalist system whether North American, European or Asian.

The objectives of this book are therefore threefold. The *first* objective is to assess the current state of knowledge in economic geography and its future directions. In doing so, this book shows how economic geographers have offered explanations of processes that affect places and lives within the broader context of the global economy. The book also offers a discussion of theoretical constructs and methodologies with a purpose to show the need to combine different approaches in understanding spatial (inter)dependencies. The *second* objective is to demonstrate the need for economic geographers to engage with multiple audiences, namely academics in different disciplines, businesses, government and non-government organizations. Within this context, this book examines how geographers have contributed to the policy-making process. One of the goals of this book is to herald a world in which economic geographers engage in conversations across disciplines (including sub-disciplines in geography) thereby creating new knowledge to promote a better understanding of processes and actions that improve lives. In the long run, the role of 'agent of change' will not be rare for an economic geographer. The third *objective* is to identify future research agenda. Contributors highlight what they see as the challenges for understanding contemporary issues, thus putting down markers for younger researchers to take the lead on.

The appeal of economic geography at the AAG conference was demonstrated by the size and diversity (students and faculty from a variety of sub-disciplines within geography from several nations) of the audience, as well as the participation from the audience, during the sessions. Contributors agreed that the impact of economic geography within and beyond geography had been constrained in the past by its own limitations. Some argued that the 'mindless' data crunching and modelling of the 1960s and 1970s that marked the 'quantitative revolution' was the beginning of the end of geography's appeal to wide audiences. Others argued that the failure to engage policymakers is another reason why economic geography, more so in the US than the UK, does not have a wide reaching influence in other social sciences or business. All recognized that both quantitative as well as qualitative methodologies are important. All argued for the need of rigour in training as we prepare a new generation of economic geographers. As a synthesizer of many disciplines and a field, which offers immense synergy in bringing together ideas and practices from other social sciences, humanities, law and business, economic geography is and should be an important component of geography pedagogy from undergraduate/freshman year through doctoral training. As many of the contributors point out, economists such as Krugman and Porter have received enormous public and academic attention and have been influential in stimulating a critical appraisal of the 'economic' within geography from within the discipline as exemplified in this book.

Through a collection of 20 chapters on theoretical constructs and methodologies, debates and discourses, as well as links to policymaking and policy evaluation, this book provides a succinct view of concepts and their historical trajectories in economic geography (see the organization of chapters below). Contributions of many other key researchers in economic geography are reflected in these chapters. The book demonstrates the differing roots and creates a common legacy in understanding dynamic dependencies in a globalized world. The contributors record changing foci and methodologies from the 1960–1980 period of quantitative economic geography, the 1980s interest in understanding how regimes of accumulation in a capitalist world construct spaces of uneven development, and how the 1990s literature was enriched by differing viewpoints and methodologies which were designed to understand the local effects of the global space economy. In the new century, especially at the Centennial Meeting of the AAG, the overwhelming response has been that of bridging gaps across 'voices within the sub-discipline of economic geography' in order to maximize our understanding of processes that shape our social, political, and economic existence. The intention of this book then is to expose its audience to the breadth of the discipline and at the same time allowing the reader to engage in current debates and understand the critical components of research in economic geography, theoretical, empirical or applied.

The book has three sections: (I) Economic Geography – Roots and Legacy, (II) Globalization and Contemporary Capitalism, and (III) Regional Competitive Advantage – Industrial Change, Human Capital and Public Policy.

In the first section, Sheppard, Hanson, McDowell, Hudson, and Scott reflect on advances in economic geography. Sheppard discusses the emergence of the field of economic geography with specific focus on the location theory, political economy, the 'cultural turn', feminist approaches, and geographical economics. At the AAG session, Eric Sheppard stated:

> Notwithstanding current frustrations with economics, the ongoing evolution of knowledge production in economic geography will necessarily continue to be shaped through its relationship to economics. That relationship is currently plagued by the shared opinion of new economic geographers in economics and geography that economics is quantitative and neoclassical, and geography is not. In fact, quantitative, non-neoclassical and post-positivist economic geography does exist, and suggests different conclusions to those dominating the new economic geography in economics. Furthermore, a variety of heterodox, post-autistic economic traditions exist (feminist, ecological, institutional, historical, Marxian, post-structural, etc.) with which new economic geographers in geography could have much in common.

His chapter explores conditions under which researchers with differing approaches can interact to strengthen knowledge production in economic geography.

Susan Hanson raises five questions: 1. What will we study? What is the domain of economic geography? 2. What are the approaches to studying economic geography? 3. What methodologies work and what don't? 4. Who are our audiences? 5. How do we teach economic geography? In answering these questions, more questions are raised on linkages across sub-disciplines of geography, the power of fieldwork and the challenge in combining multiple methodologies, as well as the need to maximize the effectiveness of economic geography in reaching out to multiple audiences, namely academics, businesses, government and non-government organizations. Pedagogic issues are explored with a full understanding of the need for evaluation through continuous dialogue among students, professors and practitioners.

Linda McDowell celebrates how feminist geographical scholarship is now mainstream. It is visible and vibrant, involving considerable numbers of scholars exploring geographies of difference and of gender relations in different parts of the world, and publishing in a range of journals. This was not always so and she records why feminist arguments were neglected and how and why academic discourses have been transformed thus the theoretical positions that lay behind the invisibility of women's lives have been dismantled. She sets out where new intellectual challenges lie for feminist geographers and how they can inform the understanding of broader audiences.

Ray Hudson traces the changing paradigms of theoretical understanding in economic geography back to radical shifts in approach in the 1950s when economic geographers returned to explaining and theorizing why economic activities are located where they are. Reviewing major advances in succeeding decades, particularly economic geography's engagement with Marxian political economy in the 1970s and its legacy, Hudson concludes that a heterodox and theoretical plural economic geography has emerged and one in which on-going debates between protagonists adhering to different theoretical positions is likely to continue. He predicts more serious theoretical engagement with relationships between economy, environment and nature.

Allen Scott's chapter fittingly completes this section with his critique of the current state of economic geography, drawing together a number of themes raised in the other chapters. He evaluates a number of prominent claims put forward in recent years by both geographers and economists about the methods and scope of economic geography. Much of his chapter revolves around two main lines of critical appraisal. He pinpoints the strong and weak points of geographical economics as it has been formulated by Paul Krugman and his co-workers. On the basis of these arguments, he identifies a viable agenda for economic geography based on an assessment of the central problems and predicaments of contemporary capitalism. This assessment leads him to the conclusion that the best bet for economic geographers today is to work out a new political economy of spatial development based on a full recognition of two main sets of circumstances: first, that the hard core of the capitalist economy remains focused on the dynamics of accumulation; second, that this hard core is irrevocably intertwined with complex socio-cultural forces, but also that it cannot be reduced to these same forces. In order to ground the line of argument that now ensues, we need at the outset to establish a few elementary principles about the production and evaluation of basic knowledge claims.

In the second section, Clark, Markusen, Walker, Daniels, Angel, Kenney and Dossani, and Yeung provide perspectives on contemporary capitalism. Gordon Clark argues that finance is the essential lens through which to study contemporary capitalism – from the local to the global. His chapter explains why and how the geography of finance is so important to the future of economic geography and how old theoretical axioms of finance are now inadequate in the light of heterogeneity of practice. Thus, he argues the need for gaining insights into new and holistic models of the structure and performance of global finance using qualitative and quantitative techniques.

Ann Markusen explores the cross-fertilization between political economy and economic geography and records major research themes from the 1970s, highlighting the advantages of the breadth of approach of economic geographers. She shows how her work and that of others on the defence industry in the 1980s has resonances for the understanding of contemporary issues. Her key concern, however, is that today's students have insufficient grounding in how the field has evolved or linking that understanding to events and movements in the larger society.

Through an account of his own experience, Richard Walker tells us about the education of an economic geographer. Economics training in the 1970s did not provide him with the answers for solving problems plaguing society. Exposure to geography at Johns Hopkins introduced him to Marxist ideology. As a junior faculty member at Berkeley, he started exploring urban topics. His initial interest in environmental issues continues and his career path shows the appreciation of diverse perspectives from the social sciences.

Peter Daniels reflects on how academic geographers have written and thought about service industries. Significant contributions came from non-geographers such as the role of service industries in economic development, uneven distribution of producer services, connections between society and the rise of the service sector among others. During the 1980s and early 1990s, economic geographers highlighted the fact that producer services are the key to understanding the basic function of services in urban or regional economies. After the mid 1990s, economic geographers examined the role of producer services in global networks and the characteristics of particular types of services. Daniels reminds us that:

> Research on the relationship between developments in information and communications technology (ICT) and the supply, demand, quality and spatial distribution services is far from exhausted, not least as offshoring and outsourcing of both routine and higher-order service tasks presents economic challenges to some developed economies and opportunities for newly emerging economies.

David Angel offers how several topics attracted enquiries in environmental economic geography: an examination of the evolution in patterns of environmental regulation of firms and industries, higher level of scrutiny of firms' activities around the world, and climate change and environmental challenges. Currently, the approaches of study of environmental economic geography include the greening of industry and the political ecology of industrial change. For the greening of industry approach, researches are firm-centred and mainly fall into three categories: consequences of changes in global production networks for economic development, technological innovations and environmental performance, and flows of capital, technology, and information and the dynamic of economic globalization. For the political ecology of industry change approach, researchs are at the beginning of theorizing the process of industrial change, which involves the flow of materials and resources as well as the flow of capital, technology, products and services. The focus is more on the structural foundation and social processes of industrial change.

Martin Kenney and Rafiq Dossani examine the potential implications of advanced telecommunications and transportation networks for the reorganization of global workforce. The recent changes in transportation and telecommunication also have an impact on services, especially high-end services like R&D. The impact on high-end services includes the need to redefine services, the relocation of services, and the unrestricted flow of digitized information. These impacts will create new labour processes; economic geographers are ideally positioned to examine the spatial implications of such processes.

Henry Yeung writes on the transformation of Asian economies. In the early 1990s, theoretical concepts in economic geography failed to capture the social and institutional contexts influencing the internationalization of firms. Peter Dicken introduced the concept of embeddedness in conceptualizing the dynamic organization of business firms. The concept has been further developed and subsequently resulted in 'business network perspective', which is utilized in explaining the economic and non-economic relations at the intra-firm, inter-firm, and extra-firm levels. Yeung has applied the above perspective to the Asian context and calls for further research in understanding the complex interrelationships in Asian capitalism.

The third section includes chapters by Martin, Asheim, Beyers, Watts, Glasmeier, Lovering, Green, and Malecki on regional competitiveness. Ron Martin's chapter focuses on the contemporary issue of regional competitiveness, with its antecedents in the pervasive phenomenon of geographically uneven development. The distinctiveness of current thinking is that disparities in performance are explicitly about competitiveness rather than 'place' competition. Martin explains discourses of competitiveness highlighting the contribution of economists, and how regional competitiveness can be seen as an evolutionary process. He argues that economic geographers have an important role to play in explaining and critiquing the idea of regional competitiveness as a way of thinking about the economic landscape and provides scope for economic geographers' engagement in public policy debates.

Björn Asheim writes about contextualizing economic geography, geography as a synthetic discipline, the co-evolution of Nordic economic geography with institutional/evolutionary economics leading to an international leading position when it comes to studies of cluster and innovation systems, and the applied side of this in accordance with the third task. He reflects on the theoretical development of the discipline, discussing the role of abstract theoretization in a Marxist tradition which was used in the early 1980s, but which seemed to have disappeared with the transition from studying Fordist to post-Fordist economic spaces. He argues that this is also related to realism as an epistemological approach, that is, the relation between abstract and concrete research, and the role played by contingencies in the economic spaces studied.

In 'Approaching research methods in economic geography', Bill Beyers reminds us that there is no one methodological or philosophical perspective that works for all. Each reader will construct for themselves their own approach and each contributor has his/her own ways of utilizing methodologies to answer research questions. Trained as a regional scientist, he values quantitative analysis, formal models, and the use of theory to frame research methods. His own research experience provides the following categorization: (1) approaches driven primarily by methods or models developed by others; (2) exploratory research motivated by pure curiosity; (3) approaches motivated by existing theories; (4) approaches driven by secondary data; (5) research driven by the development of technologies; (6) research driven by unexpected outcomes; (7) research that has value to the applied research community; and (8) collaborative research between faculty and students.

Doug Watts reflects on how the spatial organization of production within multi-regional firms has been theorized, empirically studied, and taught since the 1950s. He records that the economic landscape has changed as the contribution to regional employment by large multi-regional firms declined while the contribution of smaller firms increased from the 1980s, however, the large firm remains a key actor in the global economic system and cannot be ignored in understanding regional economic change.

Amy Glasmeier explores why economic geographers have been largely absent in policy-making circles, particularly in the USA. Her chapter records and deplores the lack of willingness for economic geographers to critically engage in debates that span academia and public policy. She argues that academic engagement in the political process best occurs when society is gravitating in that direction. Therefore at times like the 1980s, and in the current period (2006) (post-Hurricane Katrina), many streams come together in a confluence of ideas that result in an intellectual consensus about critical problems to which geographers' best efforts and significant energy are profitably aimed. She implores economic geographers not to be silent and to use their skills and knowledge vociferously engaging in debates and making a difference to how issues are being understood.

John Lovering takes to task economic geography's excessive embrace of the Empire of Capital. He provides a critique of what he identifies as the Post-Cultural-Turn Economic Geography (PCTEG), arguing that due to a number of influences, PCTEG is economic only in a thematic sense and is removed from an empirically informed awareness of the planet we live on.

Ann Green reflects on developments in labour market geographies, how they are measured and understood. Since the 1980s, methods of analysis have become more qualitative, theoretical, more focused on social and cultural issues and towards more detailed disaggregation. She identifies four major concerns of researchers: labour market adjustments, the balance between migration and commuting, area perceptions in labour market behaviour and the role of labour market intermediaries. She identifies changing policy issues and argues that economic geographers have an important role in contributing to the debate in what policy levers are available at different geographical scales to influence policy outcomes. Green argues that a central question for research is 'What is the capacity for mobility and flexibility in labour markets?'

Ed Malecki dates the late 1970s as the time when technological change was recognized by a small number of scholars, including himself, as the explanation for why companies, especially large companies, were located and how those locations changed over time. He argues that technology, broadly conceived as knowledge and application, continues to be fundamental to technological change and related regional development.

Finally, we would like to thank all of the contributors to the five sessions at the Centennial Meeting of the AAG. We are particularly grateful to Allen Scott for his keynote, which gave our event such a wonderful start. All of the sessions were enormously stimulating. We thank the audience for their participation and contribution. We hope that this book captures the excitement experienced by economic geographers, at these sessions, looking ahead at the twenty-first century.

Section I

Economic geography

Roots and legacy

1 The economic geography project

Eric Sheppard

> Genuine refutation must penetrate the power of the opponent and meet him [*sic*] on the ground of his strength; the case is not won by attacking him somewhere else and defeating him where he is not.
>
> (Adorno 1982: 5)

> The definition of what is at stake in the scientific struggle is one of the things at stake in the scientific struggle.
>
> (Bourdieu 2004: 23)

The field of economic geography,[1] a tightly focused and specialized project when I first encountered it as an undergraduate at Bristol University during the height of the 'quantitative revolution', now sprawls across several disciplines to embrace multiple theoretical, philosophical and empirical approaches. Yet, to me, at its heart has always been the goal of accounting for and redressing unequal livelihood possibilities. Explaining and redressing persistent inequalities, from place to place, in the ability of humans to pursue and attain the livelihoods that we envision for ourselves must be central to emancipatory social science. When I began, our measure of livelihood chances was straightforwardly economic and immediate (and, we would now say, developmentalist); real household incomes. It is well known that these demonstrate remarkably persistent patterns of spatial inequality from the neighborhood to the global scale, which outlive the varied modes of production envisioned to date as ways to materially underwrite society. Over time, we have become much more cautious about the adequacy of income as a measure of livelihood possibilities. Geographers now realize that unequal livelihood possibilities have to do with far more than our ability to consume. They reflect both the plethora of lifestyle choices and conceptions of the good life inhabiting the earth's surface, as well as our own conceptions of moral community – of those whose livelihood possibilities should be of concern. Economic geography has diversified accordingly.

Notwithstanding this diversification, attempts to account for geographical inequality continue to revolve around a single big question: Do capitalist economic processes (production, distribution, exchange and consumption) mitigate geographical inequalities in livelihood possibilities? This is central because of the manifest influence of capitalism over livelihood possibilities throughout the one hundred year career of economic geography. In seeking to tackle this question, economic geography has faced three further questions:

1. (How) does geography matter to the spatial dynamics of capitalism? Answers shape arguments about whether (and how) geography can contribute to our understanding of spatial inequality.
2. What is the 'economic' and what is 'geography', in economic geography? Answers shape how the big question is posed, and answered.
3. What is to be done, to redress spatial inequalities?

In the next section, I briefly caricature the remarkable diversification of theory, philosophy and method, amongst those identifying themselves as undertaking economic geography, and the diversity of answers to the above questions that has emerged. For the project of economic geography that we all contribute to, diversity can be both a strength and a weakness. In the concluding section I argue that it has been progressively more debilitating than stimulating, indicating broad schisms threatening our ability to effectively articulate a common project, but that this can and must be reversed.

Five economic geographies

Since human geographers began to take questions of theory, philosophy and method seriously in the 1960s, Anglophone economic geography (globally by far the most influential cluster of ideas) has experimented with at least five influential approaches, each with distinct perspectives on the question of geographically unequal livelihood possibilities; location theory, political economy, the 'cultural turn', feminist approaches, and geographical economics.

Location theory

August Lösch developed the radical position, for a German economist writing during the Third Reich, that market mechanisms under the rules of perfect competition could create a minimally unequal economic geographical landscape of loosely hierarchically organized central places, taking advantage of scale economies to deliver commodities at low prices (and minimal profits) to spatially dispersed rational consumers (Lösch 1954 [1940]). More than any of the initial generation of German location theorists, he offered a vision of the invisible hand operating in space that was taken up by American and British economists, geographers and regional scientists, to develop location theory in which competition organized the geography of production in a way that delivered the goods (so to speak) to consumers. In this vision, the 'economic' in economic geography meant the micro-economic (and later macro-economic) laws of economics, such as supply and demand curves and fully informed rational choice, and capitalism meant simply market exchange. Notwithstanding this grounding in economic theory, Lösch showed that geography did matter, in two ways. First, is morphogenesis; economic mechanisms can produce a spatially differentiated economic landscape, even when the geographical backcloth is undifferentiated (i.e. an unbounded uniform plain). Second, space trumps economic theory. Perfect competition is impossible on a uniform plain; rather imperfect competition prevails, with the implication that capitalists make non-zero profits (unlike the zero profits of standard microeconomic theory), reducing consumer welfare. Nevertheless, competition minimized these reductions, as well as differences between the real incomes of the most and least well off consumers (those closest and farthest, respectively, from producers).

At the macroeconomic scale, and under the assumptions of mainstream economic theory, regional scientists showed that unrestricted mobility of labor, capital, know-how and commodities also generate spatial equilibrium outcomes that tendentially minimize profits and equalize economic welfare across regions for the average consumer. Together, these results had strong normative implications. As Lösch (1954: 4) put it: 'The question of the best location is far more dignified than that of the actual location'. Inequalities in livelihood chances, defined here in terms of consumer welfare, could be reduced through the proliferation of market rationality, with the state intervening to address market failures due to the spatial nature of public goods. Philosophically, this approach aligned itself with the precepts of positivism, insisting on logical rigor and mathematical precision, and on observation as the independent arbiter of theory.

Political economy

Political economy perspectives, in the sense of Marx rather than the Milton Friedman School favored in mainstream economics, emerged in the 1970s. Early proponents had abandoned location theory, convinced that capitalist market mechanisms could never deliver the social equity that they had sought (in the furore of post-1945 social engineering). Under the echoes of 1968, these critiques galvanized a new generation of economic geographers. Just as Lösch turned to the capitalist market to redress the evils of Nazism, so Harvey (1982) turned to socialism to redress the evils of capitalism. Drawing heavily on Marx (whose arguments can be as deductive-analytical as those of mainstream economics, Roemer 1981), Harvey shows that economic inequality is inevitable because production under capitalism entails the exploitation of one class by another. Furthermore, the geography of capitalist production is bound up with uneven development and spatial divisions of labor that create geographical inequalities (e.g. dividing workers and capitalists in core regions from workers in the periphery). Capitalism is conceptualized as riven with social and geographical conflicts and contradictions that make any equilibrium at best serendipitous and temporary. Capitalism lurches from one crisis to another, with its trajectories shaped by class and spatial struggle and by the unintended consequences of economic choices and political strategies. Here, the 'economic' is centered on capitalist commodity production (the realization and accumulation of profits, and their investment in new production) rather than simply on market exchange.

Geography matters for two reasons. First, as for Lösch, space trumps economic theory. The barriers that space poses to the rapid realization of profits on capital invested in commodity production (in the form of both the built environment and the geography of communication) require modifications to Marx' theories of value, class and crisis (Harvey 1982; Massey 1984; Scott 1980; Sheppard and Barnes 1990; Webber and Rigby 1996). Second, nature constrains the imperative to accumulate and grow that is at the center of capitalist commodity production (Smith 1984). Both nature and the spatial organization of production are dialectically related to capitalism: they are shaped by, but also shape, its evolution. In this view, social movements have limited influence and unequal livelihood chances are best redressed by replacing capitalism, although little normative or empirical analysis of livelihood possibilities under more collective modes of production has been undertaken.

During the 1990s political economy came to be dominated by regulation theory. Seeking to understand capitalism's resilience, geographers sought to understand the transition from Fordism to post-Fordism and neoliberalism. Of particular interest, became the question of how localities prosper in a world where nation-states abrogate their powers to regulate their territorial economies, and investment capital is globally mobile (Scott 1998). Geography was seen to matter in two ways. Scale matters, as the nation-scale regulatory system of Fordism experienced a hollowing-out; both supra-national and sub-national scales gained in importance. Place also matters, as local political, economic and cultural conditions were seen to be crucial to economic success, although an empirical focus on success stories offered a distinctly one-sided picture. This approach has been somewhat more optimistic about the prospects of ameliorated capitalism. Methodologically, empirical work in political economy (with some exceptions) has largely privileged intensive case study research, associating quantitative and statistical methods with location theory, and with deductive, rather than dialectical thinking.

Cultural turn

The cultural turn of the 1990s, like political economy, was catalyzed by frustrations with the limitations of its forebears, combined with resonances from contemporary political and philosophical debate (about the limits of socialism, and structuralism, respectively) (Barnes 1996; Lee and Wills 1997; Thrift 1996). Initially, the cultural turn was associated with the recognition that the social and cultural contexts – within which market mechanisms are embedded – are crucial to the functionality of markets (providing legal sanction for private property, enabling economic agents to trust one another, providing moral sanction against illegal behavior, etc.), and require close analysis. Much more than context was at stake, however. It was also argued that economic processes are shaped by shared discursive understandings that make certain kinds of actions normal and others strange. How would the ideas of neoliberal globalization become so hegemonic, for example, without the ability of various right wing think-tanks to win the battle for the heart and minds of society? Furthermore, it came to be recognized that the economy consists of more than capitalist economic processes; household labor, subsistence production, LETS, the informal economy and worker cooperatives. These are undertaken in distinct places, and often are central to capitalism (reproducing its labor; cheapening labor and other inputs).

With the cultural turn, the good life is conceptualized as exceeding wealth, accumulation and development; the goals and behaviors of economic agents are not reducible to economic logic. Proponents share political economy's critique of capitalism, seeing capitalist production and exchange as facilitating rather than mitigating socially and geographically unequal livelihood possibilities. Yet they argue that political economy over-emphasizes economic mechanisms and their political consequences. Two aspects of geography are seen as important; place and networks. Socially constructed place-based practices shape context and cultural norms, some of which come to dominate by traveling beyond their local origins. Under actor-network theory, the networks connecting human and non-human actants create a distinct topological geography; a contingent relational economic geography in which scale and relative location are of diminishing importance. It is argued that unequal livelihood possibilities are best addressed by revalidating geographical difference; distinct local cultural imaginaries of the good life, and alternative economic practices.

Feminist approaches

Feminist economic geography accompanied the cultural turn, with neither reducible to the other. Beginning with feminist critiques of how mainstream economics makes invisible the extensive economic contributions of women to society that are not bought and sold in the market (50% of GDP), feminist economic geographers, while also sympathetic with the analysis and goals of political economy, argue that class is not the prime marker of livelihood possibilities. Even within the household, women often face very different livelihood possibilities from men, reflecting their distinct roles and daily geographies. Patriarchy pre-dates and exceeds capitalism, even if the forms it takes under capitalism are distinctive. As feminist theory evolved, it came to be recognized that other social markers were also vital in shaping geographical livelihood possibilities; race, age, ableness, sexuality, and location. It is argued that identity, where these markers intersect, is vital to understanding the economic actions of individuals, as it influences traditional themes in economic geography such as labor relations, workplace practices, consumption, and residential choice (Hanson and Pratt 1995; McDowell 1997). Gibson-Graham (1996) extended the feminist critique of mainstream economic theory to a far-ranging critique of political economy's failure to recognize the manifold forms of non-capitalist production that coexist under capitalism's nose. Drawing on post-structuralism, she argued that attention to and validation of such alternatives is a pre-requisite to making the possibility of a post-capitalist society real. This has catalyzed research into diverse economies. The importance of place, at a variety of scales, has also been a central theme in feminist economic geography. Diagnoses of unequal livelihood possibilities have been similar to those of the cultural turn, albeit with a greater focus on identity and on gender as a prominent marker of unequal livelihood possibilities. Methodologies have also focused on intensive case studies, but with close attention to problematizing and flattening the relationship between researcher and researched through broadly participatory research designs.

Geographical economists

Geographical economists, as a school of economic geography, emerged in the 1990s, contemporaneously with the cultural turn and feminist approaches. The catalyst came from Economics, particularly Paul Krugman (Fujita et al. 1999; Krugman 1991). In Krugman's model, firms with scale economies, each representing a different economic sector, find themselves in monopolistic competition with one another. Two regions exist, each of which produces food under constant returns, and firms choose the most profitable location. The major finding is that three different spatial equilibria are possible (agglomeration in one or the other region, or dispersed production), depending on transport costs and other parameters. There are many similarities with location theory. The 'economic' is defined in terms of what are now called 'microfoundations'; spatial patterns are an equilibrium outcome of individual, rational fully informed self-interested choices, and capitalism is reduced to market exchange. Geography matters because morphogenesis is a common outcome of such actions (agglomeration in a single region, in homogeneous space); and because (controversially, for mainstream economics) more than one spatial equilibrium is possible. Path dependence exists, in the narrow sense that different initial geographies and shifts in transportation costs determine which spatial pattern emerges. There has been extensive elaboration of this skeletal starting point, including a reinvigorated interest in constructing economic geography models within the framework of mainstream economics (Henderson and Thisse 2004). The focus on explaining the 'stylized fact' that industries agglomerate has reinforced a policy interest in place (with interesting parallels to the work of Scott within political economy, the cultural turn, and feminist approaches). It is argued that place still matters in a globalizing economy, and that fostering the competitiveness of localities as agglomerations of dynamic economic sectors is the key to both local and national economic prosperity (and by implication, for realizing livelihood possibilities).

Macroeconomic extensions of this research have returned to the old chestnut of whether the economic fortunes of regions converge under capitalist competition (answering in the affirmative, as in location theory, but using a microfoundations approach based in the new growth theory pioneered by Paul Romer). There has also been recent interest in explaining the rank-size rule. In addition, however, is an emergent interest in how physical geography affects economic growth and livelihood possibilities. Jeffrey Sachs, in particular, has undertaken research, showing correlations between economic prosperity, and both tropicality and distance from navigable waterways, arguing that physical geography matters because it acts as a barrier to the ability of market mechanisms to equalize livelihood possibilities. This has the policy implication that more global effort must be put into solving the special geographical challenges of physically disadvantaged locations. Whereas the place-based policy implications of the agglomeration school support such policies as structural adjustment (getting things 'right' locally is the key to prosperity), Sachs' work suggests that geography makes a level playing field (assumed in Krugman-like models) impossible. Methodologically, geographical economists emphasize the construction of mathematical models that produce equilibrium patterns consistent with what they identify as stylized facts (e.g. industrial agglomeration, the rank size rule, global economic inequalities), showing more interest in detailed empirical analysis and hypothesis testing than their colleagues in mainstream economics.

Strength through difference

The diversification of economic geography has resulted in a plethora of philosophical approaches, opening up an equally broad spectrum of definitions, questions and research methods. The tendency to reduce these to a series of seemingly antagonistic 'isms' belies both how they have fed off one another (in positive as well as negative ways) as well as the internal heterogeneity of each approach. Each 'ism' over-simplifies a diverse group of scholars who see themselves as engaged in a more-or-less common sub-project of economic geography. Indeed, every really existing research project inevitably draws from more than one tradition, even when the researcher claims otherwise. Figure 1 visualizes these overlaps. Crudely summarized, three distinct philosophical ideal types can be identified (the apexes), each of which has been at the center of at least one 'revolution' in economic geography since the 1950s. Each has a distinct ontology (object of knowledge) and associated epistemology, as well as a methodology for evaluating claims made about the world. For logical empiricism, sometimes loosely labeled as positivism, the object of knowledge is events experienced in the world, its epistemology (what we can know) is the identification of regular relations between observed events, and its methodology entails observation, generalization and hypothesis testing. For idealism, the object of knowledge is the models and idealizations humans impose on the world, its epistemology is to derive the meaningfulness and multiple representations of the world emanating from this, and its methodology entails hermeneutics, discourse analysis and genealogies of the emergence of and work done by these representations. For structuralism, the object of knowledge is the structures/mechanisms generating the world, its epistemology is rooted in providing theoretical accounts of these mechanisms, and its methodology is logical (variously Aristotelian and dialectical) analysis of underlying mechanisms shaping the world. As ideal types, these do not exist in their pure form. The figure suggests how various approaches heavily debated

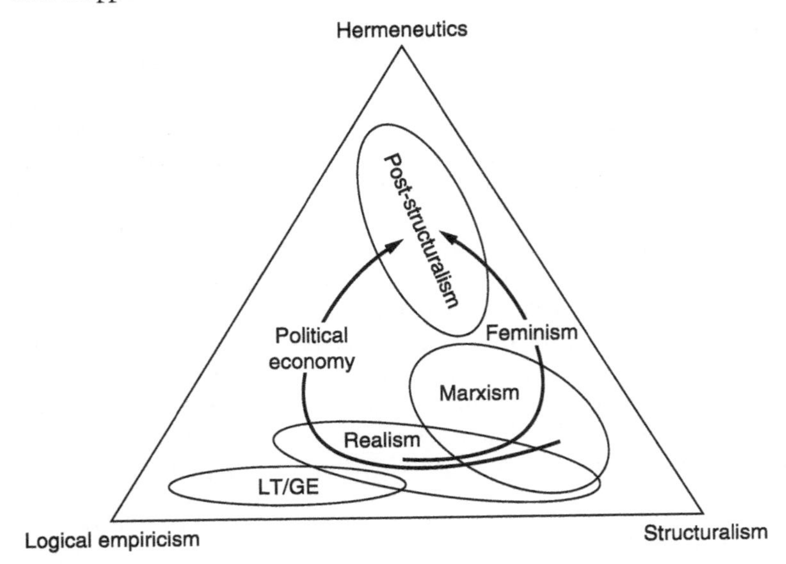

Figure 1 Economic geography's philosophical domains.

over the years in economic geography, location theory/geographical economics (LT/GE), Marxism, political economy, realism, feminism, and post-structuralism and related 'posts', range across the philosophical space bounded by these poles. In the case of feminism and political economy, clear shifts over time can also be discerned.

This diversification reflects the synthetic and trans-disciplinary nature of Geography itself, which in the case of economic geography implies taking seriously the relations between economic and other social and bio-physical processes, rather than analyzing the economic as either separable from or foundational to such other processes. Such diversity is to be celebrated, not bemoaned. Even within a discipline as apparently unified as physics, Peter Galison (1997) argues: 'It is the disorder of the scientific community – laminated, finite, partially independent strata supporting one another; is it the disunification of science – the intercalation of different patterns of argument – that is responsible for its strength and coherence.' Diversity is not necessarily regarded as a strength, however. Mainstream economics has built its reputation by excluding heterodox approaches (to the point where even Krugman's work is denigrated as insufficiently mainstream). Within economic geography, for a variety of reasons that space precludes me from detailing here, rivalry and othering of different approaches, on all fronts, has prevented us from realizing the potential of strength through difference. Two schisms, in particular, are worthy of mention; that separating approaches stressing broad theoretical claims from approaches stressing contingency and local interpretations, and that separating disciplinary cultures of Geography and Economics.

Theory vs interpretation

Influenced by feminist and post-structural philosophies, advocates of both the cultural turn and feminism have been highly critical of economic geography under the influence of either economics or Marxian political economy (Barnes 1996). In their view, these approaches make the mistake of believing in the chimerical possibility that a theory can be developed to explain much of the capitalist space economy. Notwithstanding profound differences of theory and interpretation, both approaches envision a foolproof method through which such theories can be constructed. In addition, both are overly economistic, reducing social processes to economic processes of production, distribution, exchange and consumption. By contrast, geographical economists and political economists regard the cultural turn and feminist approaches as unwilling to examine the larger picture, overly optimistic about the possibilities associated with non-capitalist economic practices, and overly focused on individual case studies (Harvey 1996; Overman 2004). Scott (2004: 491) complains: 'quite apart from its dysfunctional depreciation of the role of economic forces . . . , the cultural turn also opens the door to a disconcerting strain of philosophical idealism and political voluntarism'.

Geography vs economics

Once location theory fell out of fashion in favor of critical economic geography, geographers and economists have taken very different approaches to economic geography (compare, for example, the essays in Clark et al. 2000). For geographical economists, spatial economic structures are theorized as equilibrium outcomes of rational choices made by individual human agents. Microeconomic theory is paramount; economic value (marginal utility or productivity) is determined by market prices; production is the instantaneous conversion of inputs into outputs; livelihood possibilities are equated with income and wealth; geography is generally taken to be a given spatial structure and natural environment; and market mechanisms tendentially organize the space economy in a socially beneficial manner (with intervention necessary in the presence of market failure). Epistemologically, theories are evaluated on the basis of their ability to account for a given set of observations, and their logical (Aristotelian) rigor. For (critical) economic geographers, the capitalist economy is in a perpetual state of disequilibrium and conflict. Individual actions are constrained by (socially constructed) structures, and agents' class, gender and other social locations; capitalist economic processes cannot be separated from, and are embedded in and shaped by the social, political and cultural context; livelihood possibilities cannot be reduced to economic measures, since difference (local conceptions of the good life) must be taken into account (and cannot be reduced to preference functions); space and nature are shaped by, as well as shape, social processes; and capitalist competition tendentially reinforces geographically and socially unequal livelihood possibilities (markets are a source of social tension, not harmony).

Epistemologically, observations are taken to be theory-laden, and thus cannot be used as an independent measure of a theory's validity. By the same token, Aristotelian (mathematically deductive) logic is too narrow a framework for theory construction, as it cannot encompass dialectical mechanisms or difference. Whereas geographical economists seek to use their insights to offer policy advice so that capitalism's stakeholders can reduce the negative effects of market failure, economic geographers, while also offering short-term policy advice, in addition envision alternative economic systems with more emancipatory potential, and seek to work with marginalized members of civil society to realize such alternatives on the ground.

These differences have resulted in widespread mutual criticism. Thus economic geographers dismiss geographical economics, and often Economics in general, as overly simplistic, as pro-market, and as far too narrow – as reducing everything to economic processes. Thus Amin and Thrift (2000: 5) argue that economic geographers should 'abandon mainstream "formal" economics and take up with those pursuing economic knowledges outside economics'. Geographical economists argue that economic geographers are too negative about capitalism, lack rigor in their theoretical and empirical research, and should leave the big theoretical issues to Economics. Thus Overman (2004: 513) suggests that geographers most useful role in the division of economic geography scholarship is to contribute '[g]ood, careful case studies'.

Spaces of communication

Galison (1997: 803) suggests that Physics uses trading zones, within which ideas from different approaches can be exchanged, in order to draw strength from its diversity. 'Like two cultures distinct but living near enough to trade, they can share some activities while diverging on many others . . . in the trading zone, *despite* the differences in classification, significance, and standards of demonstration, the two groups can collaborate'. This requires certain predispositions to function: a disposition to communicate rather than exclude; institutional incentives favoring collaboration over competition; and places where trading can occur. My own experience with GIS and society indicates that under such conditions useful cooperation is possible, without reducing the exchange to one or the other approach, or to a monistic account of the world (cf. Longino 2002). Currently, however, there exist shared dualisms that prevent constructive exchange across both schisms sketched above, compounded in the latter case by power imbalance.

With respect to the theory/interpretation schism, this dualism is reinforced by a shared belief that there are only two possible philosophical approaches to making sense of the world: a foolproof method or relativism. Each is defended by one group, presenting it as superior to its other, creating either/or rather than both/and attitudes. Increasingly, however, philosophers and sociologists of science are recognizing that this dualism is indefensible, raising the possibility of exchange which is pluralist without becoming relativist (Galison 1997; Longino 2002; Bourdieu 2004). Economic geography needs to create places where, and conditions under which, non-reductionist but critical exchanges and mutual learning can occur across these contrasting epistemologies (within and between the two sides of the dualism), thereby strengthening knowledge production. We are well placed to experiment with this, because of our widespread collective experience with an epistemologically diverse context. Such conversations will require each of us to become more reflexive: to take the effort to learn other approaches from the inside and to cultivate a willingness to challenge the hard core ideas and assumptions of any approach. This is hard work: it requires the time and patience to learn about approaches that our colleagues encourage us to ignore and the courage to take unpopular positions that exceed any of the cliques we are invited to join. It also requires us to instill the same ethic in our students, at the cost of abandoning the temptation to create our 'own' schools of scholarship. In my experience, however, it has been well worthwhile.

With respect to the Geography/Economics schism, both sides share the view that Economics, defined as mainstream economics, is utterly different from Geography. In fact, however, Economics is far broader, with institutional, political economic, feminist, ecological and post-structural strands that geographers have much in common with.[2] Here, however, barriers to exchange are compounded by extreme power hierarchies, both within Economics (where 'heterodox' approaches are dismissed by the mainstream) and between (powerful) Economics and (weak) Geography. The powerful, Bourdieu (2004: 35) argues, 'enjoy decisive advantage in the competition, one reason being that they constitute an obligatory reference point for their competitors, who, whatever they do, are willy-nilly required passively or actively to take up a position in relation to them'. Bourdieu's analysis is consistent with my experience in these interactions. Even use of the (mathematical and statistical) language of geographical economics, to point out inconsistencies in its own reasoning that create space for insights from economic geographers, has resulted in responses that have left me feeling like Wittgenstein's lion: 'If a lion could talk, we could not understand him'. I thus sympathize with the frustration motivating Amin and Thrift's desire to wish Economics away, but wishing does not make it so. The ongoing evolution of knowledge production in economic geography will necessarily continue to be shaped through its relationship to Economics, a discipline with which economic geographers must remain cognizant, in all its guises, if they are to construct a more equal basis for this exchange (although, as geographers, we need to pay attention to far more than Economics).[3]

Conclusion

The shared project of economic geography offers the potential of a rich and complex analysis of the spatial dynamics of capitalism, its implications for unequal livelihood possibilities and ways to ameliorate negative implications and exceed capitalism itself. This potential cannot be found within any single approach, but in trading zones between them. A 'round table' culture can foster equal voice for the full variety of situated views on economic geography (including those from outside the Anglo-American realm, whose remarkable hegemony over the project constitutes its greatest weakness); and catalyse a culture of mutual critique and reflexive thinking, where passionate disagreement is the source of robust collective knowledge production rather than mutual alienation. Such exchanges will be simultaneously intellectual and political and will not require consensus in these domains in order to move forward (indeed consensus may be a hindrance). A pluralist non-relativist economic geography will be extraordinarily difficult to achieve and maintain, but is essential if we wish to maintain our critical edge and make a durable contribution to 'economic' analysis.

Notes

1. 'Economic geography', for the purpose of this chapter, is defined to include all researchers who identify themselves as practitioners of economic geography, irrespective of discipline. Subsequently, I will distinguish those who see this project as essentially a branch of mainstream economics ('geographical economists'), from those who see it as rooted in the social theoretic traditions that have come to dominate contemporary Anglophone human geography ('economic geographers', cf. Sheppard and Barnes [2000]).
2. Economic geographers, at the time of writing, have not attempted to engage with the lively heterodox debates in the Post-autistic Economic Review (http://www.paecon.org).
3. Too often, as was the case in the GIS and Society debates, highly simplified misrepresentations of the other side have been presented. For example, Krugman, Stiglitz and even Sachs are far from the skills for market rationality that geographers generally attribute to mainstream economists. On the other side, quantitative, non-neoclassical and post-positivist economic geography is possible.

References

Adorno, T. W. (1982) *Against Epistemology*. Oxford: Blackwell.

Amin, A. and N. Thrift (2000) 'What kind of economic theory for what kind of economic geography?', *Antipode*, 32: 4–9.

Barnes, T. (1996) *Logics of Dislocation: Models, Metaphors, and Meanings of Economic Space*. New York: Guilford Press.

Bourdieu, P. (2004) *Science of Science and Reflexivity*, trans. R. Nice. Chicago: University of Chicago Press.

Clark, G. L., M. Gertler and M. Feldman (2000) *The Oxford Handbook of Economic Geography*. Oxford: Oxford University Press.

Fujita, M., P. Krugman and A. J. Venables (1999) *The Spatial Economy: Cities, Regions and International Trade*. Cambridge, MA: MIT Press.

Galison, P. (1997) *Image and Logic: The Material Culture of Microphysics*. Chicago: Chicago University Press.

Gibson-Graham, J. K. (1996) *The End of Capitalism (as we know it)*. Oxford: Blackwell.

Hanson, S. and G. Pratt (1995) *Gender, Work, and Space*. New York: Routledge.

Harvey, D. (1982) *The Limits to Capital*. Oxford: Basil Blackwell.

Harvey, D. (1996) *Justice, Nature and the Geography of Difference*. Oxford: Basil Blackwell.

Henderson, J. V. and J-F. Thisse (2004) *Handbook of Urban and Regional Economics*. Amsterdam: North Holland.

Krugman, P. (1991) *Geography and Trade*. Cambridge, MA: MIT Press.

Lee, R. and J. Wills (1997) *Geographies of Economies*. London; New York: Arnold.

Longino, H. (2002) *The Fate of Knowledge*. Princeton, NJ: Princeton University Press.
Lösch, A. (1954[1940]) *The Economics of Location*, 2nd edn, trans. W. Stolper. New Haven: Yale University Press.
Massey, D. (1984) *Spatial Divisions of Labour: Social Structure and the Geography of Production*. London: Methuen.
McDowell, L. (1997) *Capital Culture: Money, Sex and Power at Work*. Oxford: Blackwell.
Overman, H. (2004) 'Can we learn anything from economic geography proper?', *Journal of Economic Geography*, 4: 501–16.
Roemer, J. (1981) *Analytical Foundations of Marxian Economic Theory*. Cambridge: Cambridge University Press.
Scott, A. J. (1980) *The Urban Land Nexus and the State*. London: Pion.
Scott, A. J. (1998) *Regions and the World Economy: Territorial Development in a Global Economy*. Oxford: Oxford University Press.
Scott, A. J. (2004) 'A perspective of economic geography', *Journal of Economic Geography*, 4: 479–99.
Sheppard, E. and T. J. Barnes (1990) *The Capitalist Space Economy: Geographical Analysis After Ricardo, Marx and Sraffa*. London: Unwin Hyman.
Sheppard, E. and T. J. Barnes (2000) *A Companion to Economic Geography*. Oxford: Blackwell.
Smith, N. (1984) *Uneven Development: Nature, Capital and the Production of Space*. Oxford: Basil Blackwell.
Thrift, N. (1996) *Spatial Formations*. Thousand Oaks, CA: Sage.
Webber, M. and D. Rigby (1996) *The Golden Age Illusion: Rethinking Postwar Capitalism*. New York: Guilford.

2 Thinking back, thinking ahead

Some questions for economic geographers

Susan Hanson

This book is about change: How has economic geography been changing over the past 30 years and how might it continue to change? What are the ideas and problems that have energized us as students, teachers, scholars, and practitioners of economic geography? How have the ideas and problems themselves changed (or not), and, for those questions that have persisted, how has our thinking about them changed? How can this look to the past inform the future?

Moreover, why should tomorrow's economic geographers care about what they may view as ancient history – the shaping of the discipline 30 or 40 years ago? I'm frequently asked to review for potential publication manuscripts in which the authors cite no literature whatsoever written before the dawn of the twenty-first century; others may reach as far back as 1990! As a result, such authors erroneously claim to be the first to examine *x* or *y*, they follow some of the same blind alleys visited by previous generations, and they are unable to relate their own findings to those of earlier studies and thereby build a body of scholarship on a set of questions.

As we draw upon the past to envision the future, I pose four questions in this chapter: (1) What will be the domain of economic geography, or put another way, what will count as economic geography? (2) What approaches and methods will we use? (3) What audiences will we seek to address? (4) What and how will we teach? My aim is not to provide tidy answers to these closely linked questions but rather, by thinking about each question in light of changes in economic geography over the previous several decades, to spur reflection about future directions for the field. I would also like to provoke thought about how we geographers might think constructively about the changes to come.

The only part of the future we can be certain of is the part that concerns change (we know we can count on change), but how will we shape and respond to these inevitable changes? I have seen how, over the past 40 years, geographers have managed change primarily by caricaturing, deriding, denigrating, and rejecting what has gone before and setting it in complete opposition to the preferred 'new'. (The ease with which authors fail to link their own work to earlier work seems to exemplify this point.) This approach simply does not make sense to me because it means that much of value is needlessly discredited, submerged, and lost. For me, a look at the history of this field provokes a call for greater ecumenism, for more willingness to see the connections across the decades, and for the enduring tolerance that making those connections should foster.

As I take up these four questions, a connecting thread will be a problem that has fascinated me since my graduate school days in the late 1960s and early 1970s, namely that of people's access to opportunity in the context of the geography of everyday life. The opportunities in question include, inter alia, jobs, child and elder care, health care, political participation, recreation, socializing, and shopping. This problem is one of several that have been central to economic geography for at least 40 years while having been conceptualized and analyzed differently over the years. A key dimension along which this question of access (along with many other questions) has been treated differently is that of the continuum defined by universality-particularity. Examining the various ways that people's access to opportunity has been conceptualized and investigated can serve as an example of how the past can help to inform the future. One caveat at the outset: the version of the past offered here is my understanding, my interpretation, based on my 35+ years as an academic geographer.

What will count as economic geography?

The domain of economic geography – what counts as economic geography – has enlarged in recent years. I'll first present an overview of how I see the domain of economic geography as having expanded and then use the example of access to opportunity to illustrate some of the specific ways that the field has changed.

An expanded conception of economic geography

Economic geographers have traditionally focused on production, and, within production, the emphasis until recently was on agriculture and manufacturing. Telling indicators of this emphasis were the icons on the cover of *Economic Geography* from 1950 to 1964: in one corner, a factory belching smoke; in the other corner, palm trees, a farmer and ox.[1] Another indicator is that the Economic Geography Specialty Group within the Association of American Geographers did not come into being until 1996; before then, the specialty group serving economic geographers was the Industrial Geography SG, which ceased to exist in 1996.

What counts as economic geography has expanded both within and beyond a focus on production. Within the arena of production, as the service sector has become more dominant, geographers have increasingly given more attention to services. The production lens brought to the study of services by geographers such as Bill Beyers and his students has resulted in studies exploring the location decisions of service sector firms and documenting the importance of services to regional export economies (Beyers 2005). Despite their growing willingness to encompass industries other than manufacturing, economic geographers are still wont to look at the world one industry at a time, whether the industry in question is films, computers, software, financial services, machine tools, automobiles, or retailing.

The scope of economic geography has gradually expanded beyond the realm of production, and I attribute this expansion to economic geographers' increasing interaction with urban geographers and to the blurring of the boundary between these two sub-disciplines. The traditional division of labor between economic and urban geography assigned the study of production to the economic and the study of reproduction to the urban; until the 1980s the two seemed to be separated by a firewall. In the 1960s and 1970s studies on the reproduction side (although it was not called that) focused on housing and neighborhoods, with a nod to employment only insofar as workplace location (implicitly understood to be that of the male household head) was assumed to influence residential location. Urban geographers did not pay much attention to the impact of multiple earners in households and rarely looked within the household to reveal the power relationships at work there. Economic geographers did not see economic decisions as being embedded in larger fields of social relations.

Feminist geographers, most of whom have backgrounds in urban geography, have influenced economic geography by showing the importance of the links between production and reproduction, demonstrating these ties via in-depth studies of the material circumstances of people's everyday lives in places, and thereby emphasizing the importance of place. An emphasis on place highlights the intricate and profound connections between the economic and the non-economic – indeed the difficulty of separating the two. Urban geographers, in part because of the nature of the urban, have been more comfortable than have economic geographers with the study of places, in all their confusion, complexities, and conundrums. By contrast, economic geographers have been interested in place only secondarily as it relates, for example, to industrial clusters.

In sum, what counts as economic geography has broadened both within and beyond the study of production, such that economic geographers are increasingly probing the connections between the economic and the non-economic or eroding the boundaries around what has been considered as the economic, to include the social, cultural, and political. If we want to understand how economic geographies come to be, how they function and how they change, I think we need to be alert to the interdependencies between production and reproduction and to the interdependencies created by space and place – and not just one industry at a time. I hope that what will count as economic geography will be sensitive to these interdependencies.

The question of access

For me, the really interesting questions still have to do with the relationship between people and place: How do people's actions and decisions, individually and collectively, create the structures that then enable and constrain them? The question of access – the access of people and the access of places – to opportunity has been central to economic geography for at least the past 40 years (think of Weber's industrial location model), and the changing conceptualizations of access have both reflected and driven the enlarging of economic geography described in the previous section. In general, the shift has entailed a move away from simplicity and abstraction, toward a growing appreciation for complexity and specificity in understandings of access.

From the earliest days of the so-called quantitative revolution, spatial access has been at the core of what economic geographers studied, and from the beginning, one focus has been on the relationship between human spatial behavior (individuals' movements in space) and spatial structures (the spatial organization of activity sites such as shops, workplaces, or towns). Although the dialectics and synergies of this relationship were appreciated from the start, the initial emphasis was on spatial structures as the *outcome* of spatial behaviors. An example is the body of work by Brian J. L. Berry and others aimed at empirically testing the tenets of Central Place Theory in various US contexts (Berry 1967). These studies articulated a concern for people's access to opportunities, where access was seen in terms of spatial separation (measured in distance or travel time) and opportunities were central places (agglomerations of retail facilities) of varying sizes.

With an overriding concern to tease out the important generalizations about the dependencies between human spatial behavior and the geography of retailing (with an emphasis on explaining spatial patterns), these rich empirical studies grew out of the drive to make geography a respected scientific enterprise by discovering universal laws. In fact, the empirical focus of these studies complicated Christaller's model by negating the isotropic plain and testing the extent to which actual behavior fit with the Central Place Theory assumptions of completely rational, utility-maximizing, distance-minimizing behavior.

An enduring tension in geography has been that between the general and the specific, and this tension has characterized studies of access. In particular, conceptualizations of access have come to incorporate more complex understandings of human agency as well as of geographic context. Perhaps most important, economic geographers now recognize that people sharing a common location do not necessarily share equal access to opportunities; variations among people, especially along the lines of gender, class, and race, significantly affect access (Deka 2004). Actors are no longer thought of as autonomous, independent decision makers, but rather as individuals who are embedded in various socio-geographical networks, which enable and constrain access. The geographic scales at which the contexts of access are conceptualized have moved beyond primarily the local and regional to encompass the international, as in Gerry Pratt's study of the labor markets for Filipina nannies (Pratt 1999), and the Internet and other communication technologies have further complicated the question of spatial access (Kwan and Weber 2003). For these and other reasons, the relationship between proximity and access has become more complicated.

What approaches and methods will we use?

Just as the questions posed by economic geographers have broadened in the past few decades, so have the approaches and methods that economic geographers use.

I'll briefly review some aspects of this broadening and provide an example using a study that examined access to employment.

The theoretical frameworks, approaches, and methods adopted by the leaders of the quantitative revolution in geography were largely borrowed from neoclassical economics. As a result, an approach that was widely used by proponents of spatial analysis in the 1960s entailed the development of deductive models and the subsequent testing of these models with available (secondary) data (Dacey 1966). At the core of these deductive models was economic man, that completely rational, profit-maximizing, all-knowing, a-social individual whose decisions drove the models. Economic man was the main target of behavioral geography, which emerged in the late 1960s as a reaction against the rather rigid, mechanistic form of agency implied in this neoclassical actor.

I see behavioral geography very much as an outgrowth of the quantitative revolution. For one thing, students who were inspired by this alternative to the neoclassical view of the world had been deeply immersed in mathematics, philosophy of science, mathematical modeling, and quantitative methods. The goal of behavioral geographers was not so much to reject these accoutrements of the quantitative revolution as it was to humanize economic man – to recognize that people do not have complete information, are not always distance minimizing, are embedded in networks of social relations, and therefore may base decisions on factors other than sheer economic rationality. If people do not follow the precisely prescribed decision-making calculus of economic man, how do they actually make decisions?

Although it is currently fashionable to ridicule behavioral geography as positivistic and somewhat simplistic, I see it as a highly significant phase of our disciplinary history insofar as it was the very beginning of the idea that people are different and that these differences matter immensely to decision making and to behavior. Behavioral geography represents the beginning, in geography, of looking at agency in a meaningful way by trying to grapple with some of the complexities that agency poses. From these origins have grown distinctly un-positivistic interests in, for example, difference, identity construction, meaning, positionality, and the social construction of space and scale. These more recent developments also incorporate approaches that emerged in reaction to (i.e. in order to correct some of the deficiencies in) behavioral geography, namely those approaches that pay attention to social structures, institutions, and cultures.

One example of studies of access that in some ways trace their origins to behavioral geography yet integrate other approaches is the work that Gerry Pratt and I carried out on gender and urban labor markets in Worcester, Massachusetts in the 1980s and 1990s, summarized in Hanson and Pratt (1995). This series of studies grew out of one simple finding from analyzing travel data from Baltimore, Maryland, namely that women and men who work in female-dominated occupations (such as clerical work or elementary teaching) work closer to home than do those who work in other occupations (Hanson and Johnston 1985), a finding that raised the big question that became the focus of our study: What is the relationship between geography and gender-based occupational segregation?

Through in-depth interviews with household members and employers, and through analyses that ranged in scale from intra-household to the workplace, the community, the nation state, and global movements of workers, we examined how spatial divisions of labor, and therefore (often lack of) access to opportunities, are created from the decisions of people within households and employers as well as from social networks, community structures (such as different norms and forms of child care), and local cultures (such as rootedness to place or the practice of the intergenerational transfer of housing within families).

My hope is that in their approaches and methods economic geographers of the future will combine theoretical and empirical work, will pose genuinely open-ended research questions and be open to surprises, and will productively mix qualitative and quantitative approaches. I believe it is now apparent that the scale of analysis is not related to the level of sophistication of the theoretical analysis (that is, small scale does not equate to being a-theoretical), empirical work does not equate to a lack of abstraction, and attention to complexity does not equate to disregard for theory (Massey 1994). Theory and data are closely linked, and this link can be especially productively explored via fieldwork and comparative studies. In line with my view that difference encourages creativity, I would also encourage economic geographers to engage in collaborative fieldwork with international colleagues in field sites outside one's country of origin.

I am not enthralled by studies in which the investigator knows (or at least appears to know) the answer to the research question before launching the study, and I urge students to pose research questions that are genuinely open-ended, rather than questions that are aimed at proving something the investigator is already quite certain about. Why spend several years of your life researching something you are not genuinely curious about? A related point is the challenge to be continually open to surprises throughout the research process; every open-ended research question leads to surprises, and sometimes probing these surprises is enlightening indeed. It's often interesting at the end of a project to reflect on what you learned that was truly unexpected. In this regard, a piece of advice that Professor Shalom Reichmann[2] gave me years ago was, 'Don't ignore the outliers; they can tell you a lot about what's going on'.

I am encouraged by the increasing number of economic geographers who are combining quantitative and qualitative approaches in their work, a trend that bodes well for the future. As Eric Sheppard has pointed out, there is no need to create a dualism between the quantitative and the qualitative as some are wont to do, nor is it sensible to equate one (the quantitative) with positivism and the other (the qualitative) with critical approaches in geography (Sheppard 2001). The two approaches are complementary, so that when employed together they provide a much fuller analysis than can either in isolation. Moreover, they often enable the investigator to communicate effectively with different audiences; government officials may be partial to quantitative information, for example Hanson and Pratt (2003). In using each of these approaches, however, I think we need to be as transparent as possible in describing the methods and especially the categories we use; for example, how much heterogeneity is hiding within any one category? I would also like to see us think carefully about – and raise the bar for – the standards of evidence that we collectively agree on as acceptable in economic geography scholarship, whether that scholarship uses quantitative or qualitative methods.

What audiences will we seek to address?

Far too often we fail to think about who the potential audience might be for a piece of work, or we operate on automatic pilot, such that the default audience (other academics) is the one that gets addressed. Economic geographers would do well to think ahead about the intended audience(s) for each piece of work; among possible audiences aside from our academic peers are government agencies; students, including K-12 students; private-sector groups; community groups; and non-government organizations. Of course a specific research project is likely to have multiple audiences, such that different pieces of or products from the project will be aimed at different groups. Perhaps more important, anticipating potential audiences in the early stages of planning a project can help shape the research design and, if appropriate and possible, might lead to the involvement of potential audiences in the problem definition, study design, data collection, and interpretation of results.

The benefits of involving others from such non-academic groups in the creation of knowledge they might want to put to use derive precisely from the different points of view such non-academics are likely to bring to the knowledge-creation process. While such involvement has the potential to enrich the research process, to produce knowledge that is different (from that produced purely by academics) *because* of such involvement, and to increase the use-value of the knowledge produced, it can also be a difficult process. Based on their collaborative research with community groups, Helga Leitner and Eric Sheppard describe some of the problems that such collaboration entails for both the academics and the non-academics involved (Leitner and Sheppard 2005). Bridging the often-profound differences in goals, experiences, and language is not easy. Nevertheless, economic geographers have only recently begun to engage in this kind of participatory research, and much remains to be learned about how to do it effectively.

One potential audience in particular deserves far more attention from economic geographers: the 'general public'. Economic geography is brimming with ideas and concepts that are fundamental to understanding the contemporary world. We practitioners of the field need to find ways to communicate these ideas and concepts effectively to people who lack any understanding of geography, economic or otherwise. Insofar as most of the students we teach will become members of the 'educated public' rather than practicing economic geographers, one place to start is with our teaching.

What and how will we teach?

In many ways, this last question is the most important of all; it is also where all of the prior questions come together. I would very much like to see an analysis of how, and how much, our teaching of economic geography at the university level has changed since the 1960s. Absent such an analysis I can only pose some questions for those of us who are college teachers to ponder: How might we change our courses so that we are clearly demonstrating to our students the power of geography and why, for example, economics alone does not provide satisfactory insights? Do we ask our students to connect theories in economic geography to current issues and events in the media? Do we challenge our students to become involved in solving problems in our home communities? Do we help them to see the links between local issues and global ones? Do we ask our students to get out into the field, identify researchable problems, collect and analyze data? Are we ensuring that our students, particularly at the graduate level, have a solid understanding of both quantitative and qualitative research tools? Are our students connected with local agencies and providing them with the results of student-led research?

At the regional, national, and international levels, why do we not devote more time and energy to sharing ideas about course materials, syllabi, exercises, creative assignments, successful teaching strategies and the like? Surely most of us who are academics spend a larger proportion of our time during the academic year working with students than working on our research. Why then are professional meetings so overwhelmingly focused on the presentation of research results, to the neglect of teaching and learning concerns? Perhaps the Economic Geography Specialty Group (EGSG) within the Association of American Geographers (AAG) (and similar interest groups in other national organizations) could stimulate interest in, and debate about, what we teach and how we teach; perhaps the EGSG could also coordinate a sustained exchange of ideas and materials among teachers of economic geography at the college and even the high-school level.

Students are our future. I would like to see us spend far more of our time than we currently do thinking creatively about this part of our future! There are many ways to engage with students, to develop their critical thinking and research skills, and to fire their excitement for economic geography. I don't intend to imply that we should all follow one prescribed approach in our teaching, but I do think we should have a lively exchange of ideas on this important topic.

Conclusion

A pervasive and striking theme in the chapters in this book is that change occurs not just at the level of the discipline but also at the level of the individual teacher/scholar. People's ideas do change. As obvious as this notion appears, we often seem to forget it, as when we pigeon-hole someone into one category or another. Another theme is that for most of the questions and problems that energize economic geographers today, each has a long and interesting history, a history that can and should meaningfully inform current and future work.

As I have noted elsewhere (Hanson 1992), Anglo-American geography seems to be vulnerable to violent swings of the pendulum, particularly along the axis of generality-specificity – as, for example, the swing from the high theory of environmental determinism in the 1910s and 1920s to the particularity of the regional geography of the 1930s–50s to the concern for universals in the spatial science of the 1960s and 1970s to the particularity of the post-structuralism of the 1980s and 1990s. Whereas the distinctiveness of these swings serves to sharply delineate differing paradigms, ontologies, and epistemologies, the strong rejection of previous thinking that has characterized this history is counterproductive. Instead of rejecting what has gone before (which may have more entertainment value than intellectual merit), it pays to look for common threads that can be picked up and possibly reworked to reveal new patterns and insights.

I have stressed the long intellectual history, within economic geography, of interest in people's access to opportunity. Over that history, the growing importance accorded to geographic context has increasingly called attention to the ways that access is shaped by the social relations within which individuals and groups are embedded. In addition, the complex role of information technologies in shaping access is just beginning to be investigated. The ways that differences among different types of individuals affect access are now better understood. The focus of inquiry has enlarged from examining human behavior at the individual level as the shaper of spatial structures to include examining spatial patterns and spatial structures as shapers of social processes, including those that shape access. An example of how spatial structures shape social processes is the entrepreneur who locates her firm near to a spatial cluster of a certain type of female labor (say, white, middle class, well educated; or immigrant Puerto Rican single parents) and then designs the labor process around that specific labor force. This kind of complex socio-spatial process shapes people's access to employment, and it does so in part through some of the basic principles (e.g. about people's willingness to travel) that were first articulated in the context of Central Place Theory in the 1960s.

Economic geography is interested in understanding people's livelihoods in all their complexity. In terms of what will count as economic geography, I hope that economic geographers will continue to explore livelihoods as they intersect with the wide range of social, cultural, political, and environmental processes that shape them. I hope that investigators will use a variety of approaches that allow them to combine theory and empirical work, and I hope that we will think broadly about potential audiences for our work. Most important of all, I hope that we will think as incisively and critically about teaching and learning as we routinely do about research. Our future depends on it.

Notes

1. In contrast, to signal the editors' interest in global connections, since 1992 the journal's cover has had a silhouette of Earth's landmasses.
2. Shalom was, before his death, a Professor of Geography at Hebrew University in Jerusalem.

References

Berry, B. J. L. (1967) *The Geography of Market Centers and Retail Distribution.* Englewood Cliffs, NJ: Prentice Hall.

Beyers, W. B. (2005) 'Services and the changing economic base of regions in the United States', *Service Industries Journal,* 25(4): 1–16.

Dacey, M. (1966) 'A county-seat model for the areal pattern of an urban system', *Geographical Review,* 61(4): 527–42.

Deka, D. (2004) 'Social and environmental justice issues in urban transportation', in S. Hanson and G. Guiliano (eds) *The Geography of Urban Transportation,* pp. 332–55. New York: Guilford Press.

Hanson, S. (1992) 'Reflections on American geography', *Geographical Review of Japan,* 64: 73–8.

Hanson, S. and I. Johnston (1985) 'Gender differences in worktrip length: explanations and implications', *Urban Geography,* 6: 193–219.

Hanson, S. and G. Pratt (1995) *Gender, Work, and Space.* London: Routledge.

Hanson, S. and G. Pratt (2003) 'Learning about labor: combining qualitative and quantitative methods', in A. Blunt, P. Gruffudd, J. May, M. Ogborn and D. Pinder (eds) *Cultural Geography in Practice,* pp. 106–21. London: Edward Arnold.

Kwan, M.-P. and J. Weber (2003) 'Individual accessibility revisited: implications for geographical analysis in the twenty-first century', *Geographical Analysis,* 35(4): 1–13.

Leitner, H. and E. Sheppard (2005) 'Unbounding critical geographic research on cities: the 1990s and beyond', in B. J. L. Berry and J. O. Wheeler (eds) *Urban Geography in America, 1950–2000,* pp. 349–71. New York: Routledge.

Massey, D. (1994) *Space, Place and Gender.* Minneapolis: University of Minnesota Press.

Pratt, G. (1999) 'From registered nurse to registered nanny: discursive geographics of Filipina domestic workers in Vancouver, BC', *Economic Geography,* 75(3): 215–36.

Sheppard, E. (2001) 'Quantitative geography: representations, practices, and possibilities', *Environment and Planning D: Society and Space,* 19: 535–54.

3 Feminist economic geographies

Gendered identities, cultural economies and economic change

Linda McDowell

Thinking and writing personally

It is salutary to reflect on changes in the sub-discipline, on why particular types of work and different theoretical perspectives are important at different moments. And, for those of us who have long argued against the notion of the monastic, disembodied intellectual, living on thought and air alone, much like an angel, the circumstances of everyday life that both stimulate new research questions and constrain their exploration must also enter the story. For this reason I have not only outlined changing approaches and research questions but linked them to changes in my own life. Methodological debates about reflexivity have transformed the domain of geography from the days when 'objectivity' was paramount and the personal attributes of a scholar were regarded as irrelevant. But feminist scholarship has challenged this assumption. And feminist theory and practice is what has framed my work over the years, as I was influenced by and contributed to the exciting expansion of feminist-inspired work within and beyond geography from the 1960s. From 1968, when I was a new and timid undergraduate, in different ways at different times I have continued to think, read and act within a framework largely influenced by a commitment to moving towards greater equality between men and women in the home, in the workplace and in other arenas of daily and political life. Over the intervening years there has been a remarkable shift in some of these arenas. Feminist geographical scholarship is, for example, now visible and vibrant and considerable numbers of women, and men, are involved in exploring geographies of difference, of gender relations in different parts of the world and at different times, publishing in a range of journals as well as in the specialist journal in our discipline – *Gender, Place and Culture.*

The universities have also changed over the last three decades. In Britain in 1968, about 8 per cent of women in my age group had the opportunity of going to university. Now over 40 per cent of the relevant age cohort enter higher education, and this cohort consists of as many, if not more, young women than young men, although the transfer of a large part of the costs of this education to individuals and their families is regrettable. Among the academic staff too, there is now a more equitable representation of women, although not yet among the highest ranks. But merely counting the different numbers of men and women and celebrating change is not sufficient. What is more important, at least for intellectual effort, has been the transformation of academic discourses – that wholesale critique and dismantling of the theoretical propositions that lay behind the invisibility of women's lives across the sciences, social sciences and the humanities.

As feminist scholars have argued, the ungendered notion of the rational individual in the social sciences and humanities (Pateman and Grosz 1986) and unlocated theory – what Haraway (1991) termed the 'view from nowhere' – have excluded women and women's lives from academic consideration. This view from nowhere in fact reflects the life world of the powerful and excludes daily life, the home and the politics of reproduction from the subject matter of the social sciences and the humanities on the assumption than these are merely trivial and local issues, unimportant in the grander scheme of things, than is the 'public' worlds of men. This critique is well known and largely accepted but has diffused into different sub-arenas of geography at differential rates. It has perhaps been in economic geography (and economics) that the impact of feminist scholarship and its methodological consequences has been slowest to be felt. This is not to deny the valiant efforts of a significant number of feminist economists in the US and the UK (see for example Bergman 1990; Blau and Ferber 1992; Donath 2000; Ferber and Nelson 1993; Folbre 1994; Folbre 2001; Folbre and Nelson 2000; Gardiner 1997; Humphries 1995; Jacobsen 1994; Milkman 1987; Milkman and Townsley 1994; Nelson 1992; Waring 1988) who have challenged the assumptions of their discipline and added new substantive issues, such as caring, to its agenda.

On models, Marxism and men

There are several reasons why economic geographers have, by and large, tended to neglect feminist arguments. First, like economics, the discourse and methods of economic geography are highly contested. In the 1960s, the proponents of what was then seen as a revolutionary approach – spatial science – turned against the earlier largely descriptive work on regional development – to argue for an understanding of economic landscapes based in abstract flows, best analysed by mathematical modelling, in which transportation costs and friction of distance loomed large and the resulting spatial patterns might be explained by network algebra. These disembodied, a-historical, placeless explanations had a logical elegance and a conceptual attraction that mirrors neo-classical models in economics but denied the existence of real actors, limited knowledge, vested interest and power. A response to these criticisms lay in the turn to the explanatory power of geo-historical materialism, largely stimulated by David Harvey's (1973, 1982, 1985, 1989, 1996) work and by others writing outside geography but influencing its debates. Castells' (1977) work on collective consumption and Lipietz's (1987) development of regulation theory, for example, had a huge impact on urban and economic geography respectively. Marxism, itself a contested and diverse approach, led to different emphases: on the circulation of capital, on the accumulation process, on class divisions and class struggle, providing a welcome correction to some of the absences in neo-classical and modelling approaches. But, of course, each of these approaches depends on an entirely different conception of the world, on an unbridgeable epistemological divide, still reflected in the current divisions between those who draw on versions of political economy and those who have turned to the mathematical elegance of the 'new economic geography' with its parallels in economics with the work of *inter alia* Krugman (1991) and Venables (1998). Secure within their disciplinary camps, few scholars have been prepared to redefine their theoretical or empirical work in the face of feminist critiques. Too much might be lost by giving ground in an academic environment increasingly defined by individualistic notions of competition and success.

The second reason why economic geography has been reluctant to embrace feminism is connected to the first: each of the two approaches above neglect gender relations. In the first, people seldom appear at all and in the latter, when they do figure, they are ungendered capitalists, entrepreneurs, financiers or workers and class struggle is seen as the motor of resistance. In both Castells's and Lipietz's work, for example, women's unpaid domestic labours are excluded: in the former by Castells's definition of collective consumption – the provision by the state of the services necessary to reproduce the working class – and in the latter the exclusion of women and the home from the social relations of capitalism erased the family and the home and its division of labour from examination (see McDowell 1991).

But the reasons for this neglect lie deeper in the structure of modern social thought and its distinction between the public highly regulated worlds of the economy and polity, of government, labour markets, trade and commerce and the private world of the home, constructed as a space of leisure and affection, unregulated and largely untouched by the competitive world of industrial capitalism, at least as an ideal, if seldom in practice. This division, as many feminist scholars have argued, is one that is paralleled by a gender division – the public world is a world of men, the private world that of women, based on 'natural' associations of love and care, untainted by the cash nexus. Its very naturalness needs no theoretical explanation. Here too there is a well-known critique, now generally uncontested but still ignored in 'mainstream' work. Feminist economists and geographers, however, have carefully explored the reasons for and the consequences of the absence of the private worlds of domestic labour, reproduction, caring for others, from the very definition of what is 'economic' and what activities constitute the 'economy' (Massey 1997; McDowell 1999, 2000a). Despite these arguments, some of the most thoughtful economic geographers writing today – often precisely on the contested definition of the economic (Castree 2004; Hudson 2004) and/or on the changing nature of work (Castree et al. 2004) – still ignore the gendered construction of these concepts, so neglecting large areas of unpaid and voluntary labour, largely undertaken by women.

Despite the continued neglect of these theoretical arguments by too many economic geographers, the material world has changed in recent decades in ways that mean that women's work – and so feminist arguments about its construction (for example the association of gendered traits with particular kinds of work and the under-enumeration of 'women's work' – began to loom larger on the landscape of economic geography. In the advanced industrial economies, technological change, new international divisions of labour, capital mobility and new state policies were connected with the transformation of the labour markets in these economies. Women began to be constructed and officially recognised as waged workers as well as domestic labourers in state discourses (of course women's waged work has been a key part of the economy across the centuries) and their efforts were seen as crucial to economic growth and efficiency (McDowell 2005a). From the mid twentieth century onwards, the rhetoric of women's 'dual roles' proved an acceptable way of constructing women as part-time workers, able to combine their primary role as housewives and mothers – a domestic ideology with a long tradition in Britain but which took an extreme form after the end of the Second World War, written into the postwar settlement as women's duty (Lewis 1991) – with earning a wage. Both the economy and the family seemed to prosper on this division in the postwar era as consumer industries expanded and individual families began to live in more comfortable ways, in part supported by women's ability to earn a second wage in part-time jobs.

This division of labour, however, began to flounder as deindustrialisation ripped apart the postwar compact, the families largely dependent on the wages of the male breadwinner and as new forms of service work began to expand. Here, the male-dominated trade union movement was slow to see that its complacent acceptance of lower wages for women and the development of female ghettoes in service sector employment – in the semi-professions for example of nursing, primary school teaching and social work as well as in the expanding lower echelons of service sector work – child care, retail, leisure and so on – meant that these sectors were able to be constructed as 'women's work' and so rejected as potential jobs by unemployed men who had previously worked in more masculinised forms of heavy labour. In addition these female ghettoes included mainly low paid jobs. As a result, although more and more women entered the labour market, often in casualised or part-time jobs, the overall incomes of the poorest families dependent on wage labour began to fall. Toynbee (2003) and Ehrenreich's (2001) recent exposes of the most exploitative end of these female-dominated servicing jobs in the UK and USA respectively provide shocking evidence of how the feminisation of the economy is related to growing income inequality. Meanwhile, other women – the better educated and more affluent who have benefited from the rising rates of educational participation – have begun to enter high status work in the service sector in growing numbers, moving into the professions, into law, medicine, banking, and the universities as well as into new sectors in the information economy and the cultural industries. The expansion of women's waged labour thus provided an impetus to its analysis by economic geographers, albeit not necessarily drawing on feminist explanations.

As a consequence, economic geographers interested in the transformation of their national economies, began to address new questions about regional change, organisational cultures, occupational segregation and the future shape of and spatial differentiation within service-dominated economies. And in this work, the very fact that many more workers were women, that women often worked part-time to allow for their continued responsibilities for their families, that they might not want to or be able to travel long distances to work raised new questions that had seemed irrelevant in earlier studies. Thus, for example, in a fascinating case-study of the connections between recruitment policies and locational strategies, Kristen Nelson (1987) showed how high tech firms in the Bay Area in California explicitly searched for and changed location in order to attract a certain fraction of the female labour force; reliable but unambitious middle class and middle aged women for clerical positions. Susan Hanson and Gerry Pratt (1995) in their influential study of the labour market in Gloucester Mass looked at the connections between occupational segregation, travel to work patterns and household labours, Milkman (1987) showed how local labour market conditions especially the demand for labour influenced the ways in which firms in postwar America came to different decisions about the differential pay rates between men and women. In the UK, Doreen Massey (1984) and McDowell and Massey (1984) looked at regional divisions of labour and the connections with women's domestic responsibilities. Massey (1995) then developed these ideas in an exploration of the connections between gendered roles and responsibilities in the emerging employment practices of the hi-tech industries then expanding around the university town of Cambridge and I looked at new gender divisions of labour in the City of London as deregulation after 1988 led to high demand for labour (McDowell 1997).

In all this work, what distinguished it from other studies of regional development and labour market changes was explicit attention to the causes and consequences of the gender division of labour in the workplace and in the home. Feminist economic geographers built new ways of understanding gender divisions of labour exploring why women undertake the majority of domestic labour and caring for dependents, why men and women do different jobs in the labour market, under different conditions and for different rewards, but also began to ask how assumptions about the characteristics of femininity and masculinity are themselves written into job descriptions, embedded within the cultural practices of capitalist organisations and reflected in different rates of financial remuneration.

Embodied interactive work

Over the last two decades or so, feminist theories about pleasure and desire, about the embodied performance of work as well as critiques of traditional explanations of gender segregation such as human capital theory have produced a new vocabulary and a new research agenda within (part of) economic geography. The turn to post-structuralist feminist theories as well as new work on the nature of justice has become a critical part of contemporary analyses of the new service sector and the cultural economy, in large part stimulated by the changing nature of production and work (McDowell 2000b, 2004a). Increasingly, occupations, jobs and professions in service economies are characterised by forms of work that have been defined as interactive or embodied in economies that as a whole depend on the construction and manipulation of consumer desire (Bauman 1998). One of the distinctive aspects of service sector employment lies in its very description: it is about providing a service, about servicing the needs of others, whether these needs are goods, ideas, knowledge or personal services such as a massage or a meal. In this exchange between the providers and consumers of a service, there is almost always a close personal exchange or an interaction between the provider and consumer in which the personal characteristics of the service provider take on a far greater significance than in older forms of employment and exchange. Robin Leidner (1993) has argued that in 'interactive' work in the service sector work, the bodily attributes of the service provider are an important part of the service provision. Weight, height, looks, accent and demeanour, the sexualised desirability of employees all take on an importance that was by and large irrelevant when the typical form of employment, for men at least, was in the manufacturing sector. Thus in growing numbers of service occupations, the interaction between clients and providers has become an exchange based on the manipulation of emotions and desire, a transaction in which the gender, weight, looks, bodily performance and the sexuality of the server is a key part of the exchange/seduction. Feminist philosopher, Iris Marion Young (1990) has argued that these embodied characteristics need to be part of new definitions of justice. She has explored how the idealisation of a particularly desirable body – young, slim, white – constructs various 'Others' as ineligible or less desirable workers in service economies, especially in high status forms of work.

Building on arguments about embodiment and inter-personal interactions, geographers and sociologists have begun to explore the ways in which embodied, often scripted performances in the service sector construct new patterns of inclusion and exclusion in which men and women, young and old, white and non-white workers are differently constructed and so differentially valued as potential employees. Paul du Gay (1996), for example, in a study of fashion retail outlets, has shown how a scripted exchange, based on an ideal of youthful equality, is common in clothes shops aimed at the youth market. Here the conventional distinction between the workers and clients is blurred in interactions that depend increasingly on the similarity of the sales staff and the customers and their participation in a sociable, yet scripted, ritual that is based on a false notion of equality and familiarity. In these exchanges, a groomed, trimmed, tamed and toned, sexually desirable body and the capacity for continual self-discipline is an increasingly significant aspect of the employment relationship, as it increasingly is in many professional occupations. In the retail sector, on the shop floor and elsewhere, casual flirting is a recognised part of the script in which both young men and young women, whether customers or assistants, have learnt to participate, perpetuating a myth of equality between them.

But these embodied attributes of gender – the ideal body, the maintenance of a deferential attitude to clients, the ability to seduce clients through looks and (practised) talk – are neither equally distributed nor equally maintained among the service sector workforce, and putative employers also make assumptions about prospective employees that maintain or create patterns of gender and class discrimination. Furthermore in the daily interactions and assumptions that structure relations between employees and between employees and clients the gendered body matters in different ways. In my work in the City of London I looked at the ways in which different attributes of masculinity and femininity mattered in different arenas in three merchant banks, showing how the performance of gender and the construction of a professional persona in the workplace varied both between men and women and between different spaces, but in ways that typically constructed women as 'less legitimate' employees.

What about men and boys?

Feminist work in economic geography has, as I have indicated, the potential to transform the sub-discipline. In its development, perhaps inevitably, many of the adherents of feminist perspectives, have been women, delighted to be able to write their own lives into their work and to make women's labourers visible. Similarly, many of the recent empirical analyses have fore-grounded women. However, gender is, of course, a relational concept. Women are what men are not – lack or absence in a Lacanian perspective; emotional not rational, private not public, nature to men's culture, yet, paradoxically, a gentle civilising influence on dominant, and aggressive masculinity. In the work that I have discussed so far this version of masculinity lies in the background – for example in critiques of organisational cultures that emphasis presence, long hours at the office as a mark of commitment aggression in, for example closing deals or dominating subordinates and in the work that explores different ways of managing the workplace or combining/reconciling daily life and waged work. In some studies, masculinity has been a more explicit focus. Despite being labelled as someone who works on 'women and work', my book about banking is as much about masculinity as femininity; I interviewed men as well as women about their working lives and the culture of the bank that employed them. But in a sense, it was femininity that absorbed me (indeed I re-examined my interview transcripts after a commentary from Trevor Barnes and his students that suggested I had downplayed the place of men in my work (McDowell 2001)) and certainly a set of questions about how women construct an acceptable performance in a male-dominated workplace (universities as well as banks) and combined demanding employment with raising children was a partial impetus for the study. This was a stage in my own life when my own children were still relatively young and the competing demands of home and work, publishing and nurturing seemed hard to reconcile.

The piece of empirical work that followed the bankers study also grew out of my own changing life and the questions raised by bringing up a son, but also had the explicit aim of placing men and masculinity right at the centre of the study, stimulated by an expanding set of literatures about the social construction of masculinities (see for example Connell 1995, 2000; Mac An Ghaill 1996; Whitehead 2002; Whitehead and Barrett 2001). One of the effects of the shift to a service-based economy and the associated feminisation of the nature of work has been to relatively disadvantage young working class men who, leaving school with little educational capital might, in earlier years, have expected to find relatively secure and reasonably paid work in the manufacturing sector. In most towns and cities these young men – the sort of 'lads' whose lives Paul Willis (1977) so memorably captured in his book *Learning to Labour* – now face uncertain futures as unskilled applicants for service work. The typical embodied attributes of working class masculinity – cheek, aggression, insolence, a certain style of physical presence – are no longer valued in a labour market where deference, docility and politeness is part of the scripted performance demanded in the service sector. Many of the bottom end jobs open to these men are regarded as unacceptable by them, as an insult to their sense of masculinity, as women's work. Katherine Newman (1999) in her study of fast food workers in New York showed young men found it difficult to be deferential, just as Philippe Bourgois (1995) found in his study in a New York barrio and I did in Cambridge and Sheffield (McDowell 2003). Young men disqualified themselves by their attitudes and behaviours from many of the jobs available for unskilled applicants.

Caring labour

As well as new work on gender, occupations and organisational cultures, feminist debates have placed domestic labour, both waged and unwaged, on the research agenda. Nicky Gregson and Michelle Lowe (1994), for example, wrote a splendid book, a decade ago, about the commodification of domestic labour looking at the class divisions and patterns of regional migration associated with the rise of what they termed a 'new servant class'. Placing domestic labour, especially that part of it that consists of caring for others, at the centre of economic analysis raises a further interesting set of questions about the nature of goods and services in contemporary service economies and about the principles structuring economic exchanges and how different activities are valued (McDowell 2004b). The production and maintenance of children and adults combines a number of different attributes of a service or good that are not usually recognised in the classic definitions in economics and economic geography. Care, for example, whether of children or other types of dependants, not only consists of looking after the cared-for, in the sense of making sure that no harm comes to them, but it consists of nurturing – of loving and caring for dependants and ensuring that as far as possible their well-being is secured and enhanced. Thus care is a composite good, where it is difficult to place a market value on the different aspects. Furthermore, caring is bound up with notions of love and duty, with the ideas of mutual reciprocity and is often a gift relationship outside the bounds of market exchange. Maternal love, in particular, is assumed to be 'natural', part of the social construction of femininity outlined earlier, and so such love is both beyond value and under-valued, depending on the locus of the exchange (Folbre 2001; Folbre and Nelson 2000). Typically, caring in the home, at least when the care is undertaken by a close relation to the cared-for, is unvalued. But even when the exchange takes place in the market, it is still under-valued, largely because of its association with the natural attributes of femininity and so the providers of care in the market – who are in the main women – are amongst the lowest paid workers in the labour market.

As well as the association of care with femininity, there is a further attribute of caring as an economic good that also explains its low rewards in the market. The provision of care is stubbornly resistant to productivity increases, keeping the cost of provision high despite the poor pay for employees in this sector. Care by an individual cannot easily be replaced or substituted by an alternative form of provision. It is hard to mechanise caring or to significantly extend the scope of provision and so there is little potential for economies of scale. As a consequence most care is provided in what Donath (2000) has termed 'the other economy' – provided by relatives, or through forms of reciprocal exchange, or in informal relationships – as the purchase of high quality care in the market is beyond the reach of most families. Feminist economic analysts have thus insisted on an expansion of the definition of the subject matter of their respective disciplines to include work both within the home and in the local community or in the informal sector: types of work that until recently have not loomed large in the studies of the nature of production, the allocation of labour or the rise of networked organisations in advanced industrial economies that largely constitute the subject matter of contemporary economic geography. Furthermore, the masculinist lens that defines work as waged labour in the formal economy sees only part of the question, providing a partial picture of the current transformations in the space-economy.

Conclusions: forwards not back?

The expansion of new theoretical and empirical work about gender and employment in the last few years has been exceptionally exciting and has, in my view, had a hugely beneficial impact on economic geography. As Ash Amin and Nigel Thrift (2000) have argued, economic geography used to have something of an 'anorak' image, in its tendency to dismiss power, people, difficult and contradictory lives from its remit. The desire by many economists/economic geographers to maintain their 'hard science' image has meant a continued adherence, by many, to particular rational ways of seeing the world. But in other parts of the subject, including parts of economic geography, new theoretical relationships have proved provocative and productive – with economic sociology and anthropology for example, with the cultural turn in other parts of the social sciences and the humanities as the discursive construction of identities and organisations has been explored. New issues with a wide appeal have become of a new culturally-inflected economic geography, including fashion, finance, food and sex (Amin and Thrift 2004).

I find myself, however, increasingly interested in re-thinking the past, both in a substantive empirical sense and theoretically. I have become fascinated with questions about my own past and that of other women of my own age. This seems to be a not-uncommon phenomenon as several scholars and commentators from the 'sixties' generation, including Lorna Sage (2000), Terry Eagleton (2003), Linda Grant (2002) and others, have published autobiographies or memoirs of their own upbringing and/or of their parents lives in the immediately postwar years. I too have turned to the 1940s and 1950s in a study of migrant women's working lives – in this case not of my own family but based on oral histories undertaken with Latvian women who came to Britain between 1946 and 1949 as 'volunteer' workers in the postwar reconstruction effort (McDowell 2005b). The women whom I interviewed for this study challenged my assumptions and theoretical arguments about hybridity, about multiple identities and the multiple and relational construction of the self in their insistence on the importance of an essentialised sense of national identity, as well as their position within the rigid class and gender structures of mid twentieth century Britain that constrained their lives. This work raised in a real way that set of debates that has assumed recent importance within economic geography – about how to hold together new understandings about the cultural construction of self, identity, and workplace practices with an insistence on the importance of material inequalities. As Lyn Segal (1999) has argued this debate also seems to her to be the key question in contemporary feminist scholarship.

The nature and content of economic geography have changed immeasurably since the 1960s, as has the representation of women in the labour market and women's assumptions about their future lives. New class divisions between women, and between men, have opened up in service-dominated economies as educational credentials assume growing significance in the prospects for occupational mobility and well-educated women now have more opportunities than ever before. And yet, as I have documented in my work, the structures and practices of economic institutions remain suffused with gendered assumptions and the gender divisions of labour in the home remain stubbornly inequitable, despite work/life balance policies and growing state acceptance that childcare provision is an economic issue. It may be that the enormously stimulating and challenging new research agenda in economic geography has outrun the material changes needed for a 'post-gender' world.

References

Amin, A. and N. Thrift (2000) 'What kind of economic theory for what kind of economic geography?', *Antipode*, 32: 4–9.

Amin, A. and N. Thrift (2004) *The Blackwell Cultural Economy Reader*. Oxford: Blackwell.

Bauman, Z. (1998) *Work, Consumerism and the New Poor*. Buckingham: Open University Press.

Bergman, B. (1990) 'Feminism and economics', *Women's Studies Quarterly*, 18: 68–74.

Blau, F. and M. Ferber (1992) *The Economics of Women, Men and Work*. Englewood Cliffs, NJ: Prentice Hall.

Bourgois, P. (1995) *In Search of Respect: Selling Crack in El Barrio*. Cambridge: Cambridge University Press.

Castells, M. (1977) *The Urban Question*. London: Edward Arnold.
Castree, N. (2004) 'Economy and culture are dead! long live economy and culture!', *Progress in Human Geography*, 28: 204–26.
Castree, N., N. Coe, K. Ward and M. Samers (2004) *Spaces of Work: Global Capitalism and Geographies of Labour*. London: Sage.
Connell, R. W. (1995) *Masculinities*. Cambridge: Polity Press.
Connell, R. W. (2000) *The Men and the Boys*. Cambridge: Polity Press.
Donath, S. (2000) 'The other economy: a suggestion for a distinctively feminist economics', *Feminist Economics*, 6: 115–23.
Du Gay, P. (1996) *Consumption and Identity at Work*. London: Sage.
Eagleton, T. (2003) *The Gatekeeper: A Memoir*. London: St Martin's Griffin
Ehrenreich, B. (2001) *Nickel and Dimed: On (not) Getting by in America*. New York: Metropolitan Books.
Ferber M. and J. Nelson (1993) *Beyond Economic Man: Feminist Theory and Economics*. Chicago: University of Chicago Press.
Folbre, N. (1994) *Who Pays for the Kids?: Gender and the Structures of Constraint*. New York: Routledge.
Folbre, N. (2001) *The Invisible Heart: Economics and Family Values*. New York: The New Press.
Folbre, N. and J. Nelson (2000) 'For love or money – or both?', *Journal of Economic Perspectives*, 14: 123–40.
Gardiner, J. (1997) *Gender, Care and Economics*. Basingstoke: Macmillan.
Grant, L. (2002) *Still Here*. London: Little, Brown.
Gregson, N. and M. Lowe (1994) *Servicing the Middle Classes*. London: Routledge.
Hanson, S. and G. Pratt (1995) *Gender, Work and Space*. London: Routledge.
Haraway, D. (1991) *Simians, Cyborgs and Women: The Reinvention of Nature*. London: Free Association Books.
Harvey, D. (1973) *Social Justice and the City*. London: Edward Arnold.
Harvey, D. (1982) *The Limits to Capital*. Oxford: Blackwell.
Harvey, D. (1985) *The Urbanisation of Capital*. Oxford: Blackwell.
Harvey, D. (1989) *The Condition of Postmodernity*. Oxford: Blackwell.
Harvey, D. (1996) *Justice, Nature and the Geography of Difference*. Oxford: Blackwell.
Hudson, R. (2004) 'Conceptualizing economies and their geographies: spaces, flows and circuits', *Progress in Human Geography*, 28: 447–71.
Humphries, J. (ed.) (1995) *Gender and Economics*. Cheltenham: Edward Elgar.
Jacobsen, J. (1994) *The Economics of Gender*. Oxford: Blackwell.
Krugman, P. (1991) *Geography and Trade*. Cambridge, MA: IT Press.
Leidner, R. (1993) *Fast Food, Fast Talk*. Berkeley and Los Angeles: University of California Press.
Lewis, J. (1991) *Women in Britain Since 1945*. Oxford: Blackwell.
Lipietz, A. (1987) *Mirages and Miracles: The Case of Global Fordism*. London: Verso.
Mac An Ghaill, M. (ed.) (1996) *Understanding Masculinities: Social Relations and Cultural Arenas*. Buckingham: Open University Press.
Massey, D. (1984) *Spatial Divisions of Labour*. Basingstoke: Macmillan.
Massey, D. (1995) 'Masculinity, dualisms and high technology', *Transactions, Institute of British Geographers*, 20: 487–99.
Massey, D. (1997) 'Economic/non-economic', in R. Lee and J. Wills (eds) *Geographies of Economies*, pp. 27–36. London: Arnold.

McDowell, L. (1991) 'Life without father and Ford: the new gender order of post-Fordism', *Transactions of the Institute of British Geographers*, 16: 400–19.

McDowell, L. (1997) *Capital Culture: Gender at Work in the City*. Oxford: Blackwell.

McDowell, L. (1999) *Gender, Identity and Place: Understanding Feminist Geographies*. Cambridge: Polity Press.

McDowell, L. (2000a) 'Economics, geography and gender' in G. Clark, M. Gertler and M-A. Feldman (eds) *The Handbook of Economic Geography*, pp. 497–517. Oxford: Oxford University Press.

McDowell, L. (2000b) 'Acts of memory and millennial hopes and anxieties: the awkward relationship between the economic and the cultural', *Social and Cultural Geography*, 1: 15–24.

McDowell, L. (2001) 'Men, management and multiple masculinities in organisations', *Geoforum* 32(2): 181–98.

McDowell, L. (2003) *Redundant Masculinities?: Employment Change and White Working Class Youth*. Oxford: Blackwell.

McDowell, L. (2004a) 'Sexuality, desire and embodied performances in the Workplace', in A. Bainham et al. (eds) *Sexual Positions: Sexuality and the Law*, pp. 85–107. Oxford: Hart.

McDowell, L. (2004b) 'Work, workfare, work/life balance and an ethics of care', *Progress in Human Geography*, 28: 154–63.

McDowell, L. (2005a) 'Love, money and gender divisions of labour: some critical comments on welfare to work policies', *Journal of Economic Geography*, 5(3): 365–79.

McDowell, L. (2005b) *Hard Labour: The Forgotten Voices of Latvian Migrant Volunteer Workers*. London: University College London Press.

McDowell, L. and D. Massey (1984) 'A woman's place?', in D. Massey and J. Allen (eds) *Geography Matters!*, pp. 128–47. Cambridge: Cambridge University Press.

McDowell, L., K. Ray, D. Perrons, C. Fagan and K. Ward (2005) 'Moral economies of care', *Social and Cultural Geography*, 6: 219–35.

Milkman, R. (1987) *Gender at Work: The Dynamics of Job Segregation by Sex During World War II*. Chicago: University of Chicago Press.

Milkman, R. and E. Townsley (1994) 'Gender and the economy', in N. Smelser and R. Swedburg (eds) *The Handbook of Economic Sociology*, pp. 600–19. Princeton, NJ: Princeton University Press.

Nelson, J. (1992) 'Gender, metaphor and the definition of economics', *Economics and Philosophy*, 8: 103–25.

Nelson, K. (1987) 'Labour demand, labour supply and the suburbanization of low wage office work', in A. Scott and M. Storper (eds) *Production, Work and Territory: The Geographical Anatomy of Industrial Capitalism*, pp. 96–117. London: Allen and Unwin.

Newman, K. (1999) *There's no Shame in my Game: The Working Poor in the Inner City*. New York: Vintage and Russell Sage Foundation.

Pateman C. and E. Grosz (1986) *Feminist Challenges: Social and Political Theory*. Sydney: Allen and Unwin.

Segal, L. (1999) *Why feminism?* Cambridge: Polity Press.

Toynbee, P. (2003) *Hard Work: Life in Low Pay Britain*. London: Bloomsbury Books.

Venables, A. (1998) *The International Divisions of Industry: Clustering and Competitive Advantage*. London: Centre for Economic Policy Research.

Waring, M. (1988) *If Women Counted: a New Feminist Economics*. New York: Harper and Row.

Whitehead, S. (2002) *Men and Masculinities: Key Themes and New Directions.* Cambridge: Polity Press.

Whitehead, S. and F. Barrett (eds) (2001) *The Masculinities Reader.* Cambridge: Polity Press.

Willis, P. (1977) *Learning to Labour: How Working Class Kids get Working Class Jobs.* London: Hutchinson.

Young, I. M. (1990) *Justice and the Politics of Difference.* Princeton NJ: Princeton University Press.

4 The 'new' economic geography?

Ray Hudson

Introduction

For many years, economic geography was mainly concerned with descriptive geographies of commerce, trade and Empires. However, during the 1950s some economic geographers, along with regional scientists such as Walter Isard, re-discovered the deductive location theories of von Thunen, Weber and Lösch and began again to get more seriously interested in issues of explanation and theorising why economic activities are located where they are. From the perspective of contemporary economic geography, characterised (inter alia) by a plethora of theoretical perspectives and vigorous debates as to the most appropriate forms of theory and explanation, this may seem to have been a pretty modest move forward. However, at the time it was seen as a radical and sharply contested move. More importantly, longer-term it had massive implications for the development of economic geography.

Re-focusing concerns from description of the unique to explanation of more general classes of events and spatial patterns marked a decisive and radical break. It once and for all placed the issues of explanation and theory irrevocably on the agenda of economic geography and economic geographers. This initial engagement with theory hinged on exploring the potential of mainstream neo-classically informed approaches to theorising the space-economy. Much of the next four or five decades in economic geography can be seen in terms of a series of problematisations of different theoretical positions and debates amongst their various adherents. This has involved exploring the terrain beyond mainstream economic theory, probing the links between the mainstream and various strands of heterodox economics, political economy and social theory, and acknowledging the significance of the non-economic relations that make the economy possible.

One consequence of this has been an engagement between economic geographers and a variety of other social scientists interested in the spatiality of economies – with the result that not all economic geography is carried out by economic geographers in geography departments. Far from being a problem or a weakness, this rich inter-disciplinary debate has contributed greatly to the development and intellectual vibrancy of economic geography. Reciprocally, it has been equally important in sensitising other social scientists to the significance of space in the constitution of economies and societies. In this chapter, I will briefly review the sequential emergence of a number of 'new' economic geographies, the reasons for this, and the relationships between them.

The new economic geography of the 1960s: from location theories to the behavioural geography critique of spatial science

The new economic geography of the 1960s focused attention on constructing general explanatory statements about the spatial structure of the economy as it sought to reconstruct economic geography as spatial science (for example, see Haggett 1965). Geographers sought to explain the locations of a variety of economic activities – agriculture, industry, commercial land use in cities and so on. However, the ways in which explanation was sought soon became seen to be problematic. At one level, this was because they conflated explanation with prediction; predictive accuracy became the measure of explanatory power. At another level, there were profound problems associated with an approach that sought to deduce equilibrium spatial patterns on the basis of restrictive assumptions about the natural environment, human knowledge and the character of social processes.

The fundamental difficulty was that such assumptions were indispensable to this particular deductive approach to theory building and explanation but were also both a pre-condition for and symptomatic of an impoverished and partial view of the social processes of the economy. Assumptions of the environment as an isotropic plane ignore the grounding of the economy in nature and the chronically uneven character of economic development. They also reduce the significance of spatial differentiation to variations in transport (and sometimes other production) costs within a pre-given space. Assumptions of perfect knowledge deny the fact that economic decisions are always made in a condition of partial knowledge and ignorance. Assumptions of static equilibrium deny the fact that economic processes are chronically in a state of dynamic disequilibrium, set on open-ended and unknown trajectories of change rather than inevitably and mechanistically circling around a known point of static equilibrium. In summary, while the approaches of the new economic geography of the 1960s placed questions of explanation firmly back upon the agenda of economic geographers, as a result of these limitations they did so in a way that was based upon unhelpful abstractions. Consequently, they resulted in inadequate theory, providing only weak and thin explanations that failed to grasp the essential character of the key processes that produced geographies of economies and determined the locations of economic activities.

These new approaches were soon criticised by behavioural geographers, who argued that their behavioural assumptions were untenable in an economy that exists in real space and time (for example, Pred 1967). They therefore argued the need to investigate what people actually did know, how they came to acquire this knowledge, and where they knew about, rather than assuming that they knew everything and everywhere of relevance to a particular type of behaviour. For example, behavioural economic geographers focused on the knowledge that consumers had of retail environments in order to explain who shopped where for what and that key corporate decision makers possessed about alternative locations in an attempt to explain why economic activities were located in some places rather than in others. Such approaches, built upon a partial and imperfect grasp of the relations between knowledge and the spatial organization of the economy, generally resulted in little more than descriptive accounts of behaviour, with minimal explanatory power. As such, having set out to refine an explanatory approach, behavioural geographers unfortunately fell into the descriptive trap that neo-classical location theories had set out to escape. Consequently, they quickly slipped back into obscurity but their abandonment resulted in economic geographers pushing important questions of agency from the research agenda for a decade or so.

The new economic geography of the 1970s: economic geography's engagement with Marxian political economy

The limitations of both location theories and behavioural critiques of them led economic geographers to search for more powerful conceptualisations of the processes that generated geographies of economies. In their search for more powerful explanations, economic geographers increasingly turned to Marxian political economy as a source of theoretical inspiration. Marxian political economy is centred on powerful concepts of structure, of the social structural relations that defined particular types of societies and offered a powerful challenge to the spatial fetishism of locational analysis and spatial science – that is, to the belief that spatial forms could be explained by spatial processes devoid of social content.

In the 1970s, then, economic geographers turned to Marxian political economy in order to get more powerful insights into social processes and the social grounding and relations of the economy, of what defined capitalist economies as capitalist. They recognised the need to get below and beyond the surface appearances of capitalist economies and their geographies to those structural relations and processes that had causal effectivity and that could help explain why capitalist economies and their geographies were as they were. This above all was the central issue. The concepts of value theory provided the tools to do so. Concepts such as mode of production, the dialectical class structural relationship between capital and labour, commodities and their exchange value and use value, labour-power and the labour process, and uneven development allowed a much more powerful understanding of the geographies of capitalist economies than had hitherto been possible. Extensions to include notions such as social formations and the articulation of modes of production allowed a more sophisticated understanding of the relations between capitalist and non-capitalist economies and social relations, deepening understanding of the mosaic of uneven development at multiple spatial scales. Without doubt, the most powerful and sophisticated version of this revived and enriched historical–geographical materialism emerged in 1982 with the publication of David Harvey's (1982) magisterial account of *The Limits to Capital.*

Despite subsequent critiques, economic geographers continue to argue the case for Marxian political economy. For example, Doreen Massey (1995: 307), in another of the major landmark publications of the last four decades in economic geography, *Spatial Divisions of Labour*, was at pains to emphasise the continuing relevance of Marxian political economy. For Massey, the law of value enables us to think through the broad structures of the economy and forms the 'absolutely essential basis for some central concepts – exploitation for instance'. Value theory therefore helps elucidate the social relationships specific to capitalism and its economic geographies – while recognising that there are things that value theory cannot deal with: for example, issues such as emotion and feelings cannot be captured in value categories.

In short, economic geographers continue to need Marxian political economy but they do not only need Marxian political economy. As Massey's work emphasised, specifying precisely how particular geographies of capitalist economies evolved within the structural limits defining economies as capitalist remained problematic and in turn led economic geographers to search for other approaches to theorising, either as complements to, or as alternatives to, Marxian approaches.

The new economic geographies of the 1980s: greater variety and heterogeneity

No sooner had critical geographers begun to engage with the Marxian tradition than others began to criticise them on various grounds for so doing. Whatever their specific motivation and legitimacy, however, these criticisms encouraged economic geographers to explore approaches that put more weight on agency and that allowed fuller consideration of the variety of evolutionary paths and instituted forms of capitalism through time and over space and so on. For convenience, I shall group these under three broad headings.

Agency, structure

Economic geographers exploring the potential of Marxism were accused of structural determinism, of privileging structure at the expense of (individual or collective) agency and closing off space for the effects of agency and practice, reducing people to passive 'bearers of structures'. In response to this criticism economic geographers engaged with a range of positions in modern social theory that sought to understand relationships between structure and agency. Giddens' (1984) theory of structuration (which drew heavily on the work of the geographer Torsten Hagerstrand in its approach to the time/space patterning of behaviour) was particularly influential, translated into the geographical literature by Thrift (1983). This recognised the mutually constitutive relationships between agency and structure via the social constitution of structures and the social structuration of agency, and revived interest in agency in the explanation of social action.

Within economic geography, this led to a greater attention to the knowledges, rationalities and actions of managers, workers and consumers and the ways in which these both reflected and affected their positions in the socio-spatial structures of the economy. This resulted in a more detailed understanding of how the economic geographies of capitalism evolve as a result of differential knowledges, learning processes and rationalities and the asymmetrical power relations between different groups of economic actors (for example, see Amin and Cohendet 2004; Dicken 2003; Herod 2001; Peck 1996).

Evolution, institutions, regulation

Another set of alternative approaches explored by economic geographers such as Amin (1999), Dunford (1990) and MacLeod (1997) relates to various strands of heterodox political economy and a nexus of evolutionary, institutional and regulationist approaches, often linked to theorisations of the state and to concerns with the non-economic foundations of the economy and its socio-spatial 'embeddedness'. In some cases these have clearly identifiable Marxian roots (notably strands of regulationist approaches) while in others, the origins lie more in explicitly non-Marxian approaches to political economy – which is indicative of the variety within as well as among these approaches. Irrespective of this variability, the central concern of these approaches is to elucidate the spatially and temporally variable forms that capitalist economies, their geographies and their development trajectories can take. As such, they seek to identify the variety of mechanisms and processes through which diverse capitalist economies become possible and are (re)produced.

The development of such 'middle range' theoretical concepts allows an elaboration and extension of existing concepts and ideas within Marxian political economy and a more subtle account of the historical–geographical specificity of capitalist economies. Consideration of evolutionary, institutional and regulationist approaches allows for a fuller and more nuanced elaboration of the understanding of the economic geographies of capitalism and the uneven character of capitalist development and how this is constituted. In short, and at the risk of some over-simplification, Marxian political economy, via its value theoretical approach, explains why, but such 'middle range' approaches clarify how, the uneven geographies of capitalism are constituted as they are (see Hudson 2001). From a public policy perspective, identification of these mechanisms and processes potentially allows the developmental paths of economies to be steered so as to avoid – or at least postpone – a variety of systemic crises that could threaten the accumulation process.

Culture and the 'cultural turn'

A third development relates to the recent 'cultural turn' in economic geography, with a resurgence of emphasis on cultural approaches to understanding economies and their geographies. Broadly speaking, they fall into ontological and epistemological concepts of a cultural economy (Ray and Sayer 1999).

The epistemological conception envisages the cultural as a 'bottom up' method of analysis that is complementary to a more 'top-down' political economy and focuses upon the meanings that social practices and relations have for those enmeshed in them. The ontological conception suggests that a growing culturalisation of the economy, in terms of both inputs to and outputs from it (see for example, Lash and Urry 1994), has led to the economy becoming ontologically more cultural.

It is certainly true that in some respects economic practices have become more sensitive to cultural differences. Corporations are increasingly aware of this and indeed have helped promote it via their advertising and brand management strategies and in other ways (for example, in representing work as the route to self-fulfilment and personal development) as part of capital's enduring concern to raise productivity, increase sales and speed up the pace of accumulation more generally. To some perhaps considerable extent, this 'cultural turn' in corporate practice reflects a growing concern with the knowledge base of the economy and the ways which the economy is thought about and talked about, issues that have increasingly come to interest business consultants, academics in Business Schools and some influential economic geographers (for example, Thrift 2005). However, while some economic geographers have embraced this 'cultural turn', others remain much more sceptical about claims that the economy has become ontologically more cultural and caution against the dangers of conflating changes in academic fashion with changes in the economy and its practices.

Rather than approach cultural economy and political economy as an either/or choice, some economic geographers have attempted to forge a synthesis and develop a culturally sensitive political economy that begins from the assumption that the economy is – necessarily – always cultural but one that is always alert to the materialities, power geometries and dynamics of political economy (Hudson 2005). Such an approach to cultural political economy can be further developed by exploring the constitutive role of semiosis – that is, the inter-subjective production of meanings – in economic and political activities and institutions and the social order more generally. This leads Bob Jessop (2004) to argue that cultural political economy is a 'post-disciplinary' approach that adopts the 'cultural turn' in economic and political inquiry without neglecting the articulation of semiosis with the inter-connected materialities of economics and politics within wider social formations.

Re-thinking and problematising the economy

A further twist to the evolving tail/tale of economic geography, linked to the growth of interest in cultural economy approaches, is that there has been an increasing concern with the conceptualisation and definition of what we as economic geographers take to be 'the economy'. In part this is rooted in older concerns, such as those of feminists and/or Marxists as to the conceptualisation of domestic labour and unpaid work in the home that is critical to the reproduction of labour-power in capitalist economies, in part it is related to more recent post-structuralist concerns with deconstructing the economy (for example, see

Gibson-Graham 1996). It has also become linked to interests in 'alternative' economies that exist on the margins of, or in the interstices of, the mainstream capitalist economy (for example, see Leyshon et al. 2003). This is important in creating space for imagining alternative forms and spaces of economic relations and theorisations of 'the economy' and its geographies.

There are however dangers, as Scott (2004: 491) has recently emphasised in relation to Gibson-Graham's (1996: 206) announcement that 'the way to begin to break free of capitalism is to turn its prevalent presentations on their head'. As he acerbically points: 'Presto. . . . The claim is presented in all its baldness, without any apparent consciousness that attempts to break free of any given social system are likely to run into the stubborn realities of its indurated social and property relations as they actually exist.' In arguing for a serious consideration of culture but against the 'cultural turn', Scott goes on to suggest that 'quite apart from its dysfunctional depreciation of the role of economic forces and structural logics in economic geography, the cultural turn also opens a door to a disconcerting strain of philosophical idealism and political voluntarism in modern geography'. But it is precisely such economic forces and structural logics that shape the often brutal economically dominated world that economic geographers need to be able to grapple with and understand.

What goes round comes round: the new geographical economics/new economic geography of the 1990s

In the 1990s, a number of prominent economists, perhaps most prominently Paul Krugman, began to acknowledge the importance of space in the constitution of the economy and to alter the approaches of mainstream economics to allow for this. Not surprisingly, this proved to be attractive to many economic geographers who had previously felt neglected and ignored by mainstream economists and who seized the opportunity to engage in debate with them (see for example, Clark et al. 2000) and equally worrying to others (see Amin and Thrift 2000 and the subsequent debates in the pages of *Antipode* 2001).

It is worth emphasising that the critique developed of 1960s location theories remains substantially valid for the New Economic Geography and the New Geographical Economics of the 1990s, which remain essentially committed to methodological individualism and thinly socialised explanatory accounts. Strictly speaking, Krugman's work and the work of others that it has inspired are not neo-classical, firmly eschewing any notion of constant returns to scale and perfect competition. However, that said, it retains a strong kinship with mainstream economics by reason of its commitment to methodological individualism, full information utility-maximising individuals and profit-maximising firms, and an exclusive focus on socially disembedded relations of exchange. Indeed the rise of technically more sophisticated versions of the neo-classical location theory orthodoxies of the 1950s and 1960s is indicative and symptomatic of an attempt to revive approaches that were then revealed as seriously flawed and limited in their explanatory power and sophistication. As such, it raises interesting questions as to why such approaches are making a comeback of sorts in economic geography (which I turn to briefly in the next concluding section of the chapter).

Conclusions

Economic geography is now characterised by a plethora of sometimes competing, sometimes complementary, sometimes simply indifferent to one another, theoretical positions. This is both exciting and testimony to the intellectual vigour of economic geography. However, it is also potentially confusing, especially for those seeking to pick their way through these entangled positions as they are grounded in different assumptions as to what is important, what matters and what economic geography ought to be – and it is important to keep in mind this normative dimension. To some extent, the developmental trajectory of economic geography can itself be understood in terms of instituted behaviour and path dependent development, with what was once seen as new becoming seen as old and with the (allegedly) new sometimes involving a re-discovery and re-invention of the old. However, the end-product of this complex and emergent process is a heterodox and multiply theoretical economic geography, which certainly accommodates a wide variety of viewpoints and provides the arena for vigorous discussion as to their varying merits.

How should we explain this complex trajectory and proliferation of co-existing theoretical positions? In short, we can identify three sorts of reasons. First, a genuine concern with the explanatory limitations of particular theoretical positions, arising from the perception of conceptual lacunae within them and an exploration of other theoretical positions that allow these perceived weaknesses to be better addressed. Second, the impacts of generational disciplinary sociologies, as newly qualified economic geographers felt the need to carve out and define their own intellectual territory, to 'do something different' from the existing orthodoxies. This is neither surprising nor necessarily unwelcome. After all, the engagement of differing and conflicting ideas is what drives forward understanding and theory, as long as – and this is a key caveat – this is not simply a fashion effect and change for the sake of change. Third, there were the effects of political correctness and 'insidious careerism' (Walker 1989: 151), which were particularly evident in some critiques of and moves away from Marxian political economy. This was particularly so with the rise of neo-liberalism from the 1970s, for there is no doubt that for some being seen to be Marxist was perceived as a career threatening move while for others having a go at Marxism was one way of establishing one's credentials as a safe bet. No doubt these – and other – motives became entangled in particular ways in particular cases, but the aggregate end result is clearly visible in the multiplicity of positions observable in contemporary economic geography.

What then of the future of economic geography, theoretically and substantively? Given the present position, I think it certain that economic geography in the foreseeable future will continue to be characterised by a theoretical plurality and on-going debates between protagonists adhering to different theoretical traditions. I take this to be a good thing for the economic geographies of the late modern capitalist world are too complex and nuanced to be explicable simply in terms of one all-encompassing theoretical position. What might the next 'new' economic geography substantively focus on? I think that there is a very fair chance that this will involve a more serious engagement with issues of nature, the materiality of economic processes and their 'environmental footprint' and a growing theoretical attention to relationships between economy, environment and nature. In saying this, I acknowledge that there have been important contributions on these issues by economic geographers but the growing threat of global warming and other forms of ecological change induced by the economic practices of people will force these issues more prominently onto the agendas of economic geographers – among many others.

References

Amin, A. (1999) 'An institutionalist perspective on regional development', *International Journal of Urban and Regional Development*, 23: 365–78.

Amin, A. and P. Cohendet (2004) *Architectures of Knowledge*. Oxford: Oxford University Press.

Amin, A. and N. Thrift (2000) 'What kind of economic theory for what kind of economic geography?', *Antipode*, 32: 4–9.

Clark, G. L., M. P. Feldman and M. S. Gertler (eds) (2000) *The Oxford Handbook of Economic Geography*. Oxford: Oxford University Press.

Dicken, P. (2003 [1986]) *Global Shift*, 4th edn. London: Sage.

Dunford, M. (1990) 'Theories of regulation', *Society and Space*, 8: 297–321.

Gibson-Graham, J. K. (1996) *The End of Capitalism (as we knew it)*. Oxford: Blackwell.

Giddens, A. (1984) *The Constitution of Society*. Cambridge: Polity.

Haggett, P. (1965) *Locational Analysis in Human Geography*. London: Arnold.

Harvey, D. (1982) *The Limits to Capital*. Oxford: Blackwell.

Herod, A. (2001) *Labor Geographies*. New York: Guilford.

Hudson, R. (2001) *Producing Places*. New York: Guilford.

Hudson, R. (2005) *Economic Geographies: Circuits, Flows and Spaces*. London: Sage.

Jessop, B. (2004) 'Critical semiotic analysis and cultural political economy', *Critical Discourse Studies*, 1: 159–74.

Lash, S. and J. Urry (1994) *Economies of Signs and Space*. London: Sage.

Leyshon, A., R. Lee and C. C. Williams (eds) (2003) *Alternative Economic Spaces*. London: Sage.

Macleod, G. (1997) 'Globalizing Parisian thought waves: recent advances in the study of social regulation, politics, discourse and space', *Progress in Human Geography*, 21: 530–53.

Massey, D. (1995 [1984]) *Spatial Divisions of Labour*, 2nd edn. London: Macmillan.

Peck, J. (1996) *Work-Place*. New York: Guilford.

Pred, A. (1967) *Behaviour and Location*. Lund: The Royal University of Lund.

Ray, L. and A. Sayer (1999) 'Introduction', in L. Ray and A. Sayer (eds) *Culture and Economy After the Cultural Turn*, pp. 1–16. London: Sage.

Scott, A. J. (2004) 'A perspective on economic geography', *Journal of Economic Geography*, 4: 479–99.

Thrift, N. (1983) 'On the determination of social action in time and space', *Society and Space*, 1: 23–57.

Thrift, N. (2005) *Knowing Capitalism*. London: Sage.

Walker, R. (1989) 'What's left to do?', *Antipode*, 21: 133–65.

5 A perspective of economic geography

Allen J. Scott

In search of perspective

In this chapter, I attempt to evaluate a number of prominent claims put forward in recent years by both geographers and economists about the methods and scope of economic geography. Much of the chapter revolves around two main lines of critical appraisal. First, I seek to highlight the strong and weak points of geographical economics as it has been formulated by Paul Krugman and his co-workers (though I also acknowledge that geographical economics is now moving well beyond this initial point of departure). Second, I provide a critique of the version of economic geography that is currently being worked out by a number of geographers under the rubric of the cultural turn, and here I place special emphasis on what I take to be its peculiar obsession with evacuating the economic content from economic geography. On the basis of these arguments, I then make a brief effort to identify a viable agenda for economic geography based on an assessment of the central problems and predicaments of contemporary capitalism. This assessment leads me to the conclusion that the best bet for economic geographers today is to work out a new political economy of spatial development based on a full recognition of two main sets of circumstances: first, that the hard core of the capitalist economy remains focused on the dynamics of accumulation; second, that this hard core is irrevocably intertwined with complex socio-cultural forces, but also that it cannot be reduced to these same forces. In order to ground the line of argument that now ensues, we need at the outset to establish a few elementary principles about the production and evaluation of basic knowledge claims.

A large recent body of work in the theory of knowledge and social epistemology has made us increasingly accustomed to the notion that research, reflection, and writing are not so much pathways into the transcendental, as they are concrete social phenomena, forever rooted in the immanence of daily life. By the same token, knowledge is in practice a shifting patchwork of unstable, contested, and historically-contingent ideas shot through from beginning to end with human interests and apologetic meaning (Barnes 1974; Latour 1991; Rorty 1979; Shapin 1998). Mannheim (1952), an early exponent of the sociology of knowledge, expressed something of the same sentiments in the proposition that the problems of science in the end are mediated outcomes of the problems of social existence.

Postmodernists, of course, have picked up on ideas like these to proclaim the radical relativism of knowledge and the dangers of 'totalization' (cf. Dear 2000), though the first of these claims carries the point much too far in my opinion, and the second turns out on closer examination to be largely a case of mistaken identity. I accept that knowledge is socially constructed and not foundational, but not that it is purely self-referential, for although knowledge is never a precise mirror of reality, it does not follow – given any kind of belief that some sort of external reality actually exists – that one mirror is as good as another (Sayer 2000). The aversion to so-called totalization among many geographers today seems to translate for the most part, in a more neutral vocabulary, into the entirely sensible principle that theories of social reality should not claim for themselves wider explanatory powers than they in fact possess. However, the principle strikes me as pernicious to the degree that it is then used to insinuate that small and unassuming concepts are meaningful and legitimate whereas large and ambitious concepts are *necessarily* irrational. This in turn has an unfortunately chilling effect on high-risk conceptual and theoretical speculation.

These brief remarks set the stage for the various strategies of assessment of economic geography that are adopted in what follows. We want to be able to account for the shifting substantive emphases and internal divisions of the field in a way that is systematically attentive to external contextual conditions, but which does not invoke these conditions as mechanical determinants. We must, in particular, be alert to the social and institutional frameworks that encourage or block the development of ideas in certain directions, as well as to the professional interests that drive choices about research commitments. Moreover, since science is (either consciously or unselfconsciously) a vehicle for the promotion of social agendas, we need to examine the wider ideological and political implications of any knowledge claims. A basic question in this regard is: whose interests do they ultimately serve, and in what ways? The simple posing of this question implies already that the form of appraisal that follows entails a degree of partisan engagement (Haraway 1991; Yeung 2003), though in a way, I hope (given my preceding critical comments on relativism), that maintains a controlled relationship to an underlying notion of coherence and plausibility. Last but by no means least, then, we must certainly pay close attention to the logical integrity, the scope of reference, the correspondence between ideas and data, and so forth, of the various versions of economic geography that are on offer.

Economic geographers at work

Geography and the disciplinary division of labor

Geographers long ago gave up trying to legislate in a priori terms the shape and form of their discipline. In any case, from what has gone before, we cannot understand geography, or any other science for that matter, in relation to some ideal normative vision of disciplinary order. Geography as a whole owes its current standing as a distinctive university discipline as much to the inertia of academic and professional institutions as it does to any epistemological imperative. The geographer's stock-in-trade, nowadays, is usually claimed to revolve in various ways around questions of space and spatial relations. This claim provides a reassuring professional anchor of sorts, but is in practice open to appropriation by virtually any social science, given that space is intrinsically constitutive of all social life. In fact, geographers and other social scientists regularly encounter one another at points that lie deep inside each other's proclaimed fields of inquiry, and this circumstance reveals another of modern geography's peculiarities, namely its extreme intellectual hybridity. It is perhaps because of this hybridity that geography is so susceptible to rapidly shifting intellectual currents and polemical debate, but also – and this is surely one of its strengths – an unusual responsiveness to the burning practical issues of the day.

The wayward course of economic geography in the last half-century

Economic geography reproduces these features of geography as a whole in microcosm. On the one side, it is greatly influenced by issues of social and political theory. On the other side, given its substantive emphases, it has particularly strong areas of overlap with economics and business studies. At any given moment in time, it nevertheless functions as a more or less distinctive intellectual and professional community that brings unique synthetic perspectives to the tension-filled terrain(s) of investigation that it seeks to conquer. At the same time, economic geography has been greatly susceptible to periodic shifts of course over the last several decades, often in surprising ways, and equally often with the same dramatis personae, as it were, appearing and re-appearing in different costumes in different acts of the play.

The period of the 1950s and 1960s was especially important as a formative moment in the emergence of economic geography as a self-assertive subdiscipline within geography as a whole. This was a period of great intellectual and professional struggle in geography between traditionalists and reformers, with the latter seeking to push geography out of its perceived idiographic torpor and – on the basis of quantitative methodologies and formal modeling – into a more forthright engagement with theoretical ideas (Gould 1979). Economic geographers were in the vanguard of this movement, and they were able to push their agendas vigorously, partly because of their strategic affiliation with a then-powerful regional science, partly because the questions they were posing about the spatial organization of the economy were of central concern to much policy making in the capitalism of the era, with its central mass-production industries and its activist forms of social regulation as manifest in Keynesian economic policy and the apparatus of the welfare state (Benko 1998; Scott 2000).

This early moment of efflorescence was succeeded by a sharp turn toward political economy as the crises of the early 1970s mounted in intensity, and as the general critique of capitalism became increasingly vociferous in academic circles.

This was a period in which geographers developed a deep concern about the spatial manifestations of economic crisis generally, as reflected in a spate of papers and books on topics of regional decline, job loss, regional inequalities, poverty, and so on (Bluestone and Harrison 1982; Carney et al. 1980; Massey and Meegan 1982). It was also a period in which much of economic geographers' portrayal of basic social realities was cast either openly and frankly in Marxian terms or in variously marxisant versions. The first stirrings of a vigorous feminist encounter with economic geography also began to take shape at this time.

As the initial intimations of the so-called new economy made their appearance in the early 1980s, and as the crisis years of the 1970s receded, economic geography started to go through another of its periodic sea changes. A doubly-faceted dynamic of economic and geographic transformation was now beginning to push geographers toward a reformulated sense of spatial dynamics. On the one hand, new spatial foci of economic growth were springing up in hitherto peripheral or quasi-peripheral regions in the more economically-advanced countries, with neo-artisanal communities in the Third Italy and high-technology industrial districts in the US Sunbelt doing heavy duty as early exemplars of this trend (Becattini 1987; Scott 1986). In this connection, geographers' interests converged intently on the theoretical and empirical analysis of spatial agglomeration. On the other hand, a great intensification of the international division of labor was rapidly occurring, especially under the aegis of the multinational corporation (Fröbel et al. 1980). In this connection, the main issues increasingly crystallized around globalization and its expression in international commodity chains, cross-border corporate linkages, capital flows, foreign branch plant formation, and so on (e.g. Dicken 1992; Johnston et al. 1995; Taylor et al. 2002). The themes of agglomeration and international economic integration more or less continue to dominate the field today, though many detailed changes of emphasis have occurred as research has progressed. Indeed, of late years, these two themes have tended increasingly to converge together around the notion of the local and the global as two interrelated scales of analysis within a process of economic and political rescaling generally (Swyngedouw 1997).

These thematic developments represent only a thumb-nail sketch of the recent intellectual history of economic geography. We must recognize that there have been many additional twists and turns within this history, both of empirical emphasis and of theoretical debate. As it stands, however, this account now serves as a general point of entry into a detailed examination of some of the major conceptual tensions that run through the field today, including a number of claims, which if they can be sustained, presage some quite unexpected new directions of development.

Turbulence and challenge

Economic geography, then, has been marked over its postwar history by a great susceptibility to turbulence. A notable recent sign of this tendency is the various 'turns' that the field is said to have taken or to be about to take. A cursory count reveals an empirical turn (Smith 1987), an interpretative turn (Imrie et al. 1996), a normative turn (Sayer and Storper 1997), a cultural turn (Crang 1997), a policy turn (Martin 2001), and a relational turn (Boggs and Rantisi 2003), among others.

In some instances, the proclamation of these turns has been no more than an attempt to test the waters. In others, it has registered some real underlying tendency in geographic research. Today, the field is subject to particularly strong contestation from two main sources. One of these lies largely outside geography proper and is being energetically pushed by economists under the rubric of a new geographical economics. It represents a major professional challenge to economic geographers by reason of its threatened appropriation and theoretical transformation of significant parts of the field. The other is represented by the cultural turn that comes in significant degree from within geography itself, but also reflects the wider politicization of cultural issues and the rise of concerns about identity in contemporary society. The cultural turn represents a very different kind of challenge to economic geography on account of its efforts to promote within the field a more highly developed consciousness of the role of culture in the eventuation of economic practices. Much of the rest of this chapter is concerned with investigating the nature of this current conjuncture in the light of the arguments already marshalled.

Geographical economics: accomplishments and deficits

The core model

Krugman's *Geography and Trade,* published in 1991, rang a tocsin in the ears of geographers, with its twofold proclamation that the project of economic geography was now at last beginning, and that economic geographers (of the variety found in geography departments) had hitherto been more or less sleeping at the wheel.

The new geographical economics did not, as we might expect, reach back to regional science, but appeared quite unexpectedly from another quarter: the new growth and trade theories that had been taking shape in economics over the previous decade or so (Meardon 2000; Thisse 1997). The core model is built up around the idea of monopolistic competition as originally propounded by Chamberlin (1933) and subsequently formalized by Dixit and Stiglitz (1977). The model also has some points of resemblance to an older tradition of heterodox economics focused on increasing returns and cumulative causation, as represented by Hirschman (1958), Myrdal (1959), and Kaldor (1970). Strictly speaking, Krugman's model, and the surge of research activities that it has sparked off, are not neoclassical, for it firmly eschews any notion of constant returns to scale and perfect competition. That said, the model retains a strong kinship with mainstream economics by reason of its commitment to methodological individualism, full information, utility-maximizing individuals and profit-maximizing firms, and an exclusive focus on socially disembedded relationships of exchange (Dymski 1996).

The model itself is an ingenious if convoluted piece of algebra. Imagine a set of regions[1] with production represented by immobile farmers and mobile manufacturing workers and firms. Manufacturing firms engage in product differentiation (monopolistic competition) with increasing returns to scale, or better yet, unexhausted economies of scale. Thus, each firm produces a unique or quasi-unique variety in its given product class. Consumers in all regions (both farmers and manufacturing workers) purchase some portion of every firm's output. Wages are determined endogenously. Market prices always reflect the transport costs incurred in product shipment. Consumers in regions with many producers will therefore pay less than those less favorably situated. Any individual's 'utility' is a function of both nominal wages and price levels. Mobile manufacturing workers will migrate from (peripheral) regions with lower utility to (core) regions with higher utility. Nominal wages in any region whose manufacturing labor force is increasing in this way will tend to fall (though corresponding utilities will increase because of the decreasing cost of final goods). More and more manufacturing firms will therefore be attracted to the region, which will in turn induce further in-migration of labor. The net result will be a path-dependent process of spatial development leading to a stable core-periphery pattern. Eventually an equilibrium of production, wages, prices, and demand will be attained, and the final result will exhibit market-driven pecuniary externalities (i.e. overall real price reductions) derived from intra-firm increasing returns under conditions of Chamberlinian competition. In a later formulation of the core model, Krugman and Venables (1995) showed that core-periphery contrasts will tend to be relatively subdued (or even to disappear entirely) in situations where transport costs are uniformly very high or very low, whereas core-periphery contrasts will be maximized where transport costs are contained within some intermediate range of values.

Depending on the distribution of immobile workers, transport costs, elasticities of demand and substitution, and other basic parameters, the model is capable of generating widely varying locational outcomes. Numerous modifications and extensions of the basic model have been proposed since its first formulation. For example, Krugman and Venables (1995, 1996) and Venables (1996) introduce inter-industrial linkages into the model. Abdel-Rahman and Fujita (1990) have suggested that the production functions of downstream industries are sensitive to the variety of available upstream inputs. In this case, agglomeration and pecuniary externalities are brought about by the productivity effects of input variety. In another variation of the model, Baldwin (1999) has shown how demand-linked circular causality can induce agglomeration, even in the absence of labor mobility.[2]

The geographers' reception of geographical economics

Rather predictably, the geographers' first reaction to the new geographical economics was one of virtually unqualified rejection. In a brief review of Krugman's book, Johnston (1992: 1066) dismisses it with the comment 'not recommended'. Martin (1999: 67) writes that geographical economics, 'is not that new and it most certainly is not geography'. In a similar vein, Lee (2002: 353) rebuffs the entire enterprise with the comment that 'there are . . . precious few grounds for some mutually beneficial conversation here'.

Something of these reactions can no doubt be ascribed to geography's endemic professional anxieties reflecting its relatively low standing on the academic totem pole, certainly by comparison with economics. Krugman's tasteless self-promotion as the creative genius par excellence of economic geography, and the champion of four-square thinking generally, did nothing to assuage those anxieties. The geographers' main complaints about this work have tended to revolve around their concern that it is unduly cut off from the wider social and political frameworks within which economic issues are actually played out. Many economic geographers' perception of their own work, as well, points to practices of research that are grounded, open, polycentric, focused on rich empirical description, and deeply conscious of the contingency and complexity of things (Boddy 1999; Thrift and Olds 1996). I shall take issue with this particular line of self-justification later, but let us for the moment simply note some of its basic modulations. Thus, Clark (1998: 75) suggests that 'a fine-grained substantive appreciation of diversity, combined with empirical methods of analysis like case studies are the proper methods of economic geography'. Martin (1999: 77) castigates geographical economics for its neglect of 'real communities in real historical, social and cultural settings'. In a similar vein, Barnes (2003), picking up on the work of Geertz (1983), proclaims that all knowledge is local and that locational analysis in geography is (or should be) born out of specific contextual settings.

Some of these comments point to significant deficiencies of geographical economics, and they need to be taken seriously (see also David 1999). In my opinion, however, they provide at best only peripheral glosses on the main issues at stake and they fail signally to grapple with the target's central weakness, which, as I hope to demonstrate, reside in its limited *analytical* grasp of agglomeration economies and locational processes. I would also argue that the geographers' critique has tended to veer too enthusiastically in favor of the virtues of the empirical and the particular and too forcefully against theoretical systematization and formal analysis, thereby implicitly abdicating from far too much that is of value on their own side (though I suspect that most of the geographers mentioned earlier would not consider this to be a fair judgment of what they are saying). In any case, a scientifically meaningful and politically progressive economic geography can scarcely allow itself to be reduced merely to close dialogue with endless empirical relata (Plummer and Sheppard 2001; Sayer 2000). A more penetrating engagement with the internal theoretical structure of geographical economics, it seems to me, is more than overdue.

An evaluation of the Krugman model

One of the obvious failures of earlier neoclassical theories in economic geography and regional science is that their commitment to perfect competition and constant returns induced them to overemphasize the divisibility of economic activities, leading in turn to a radical underemphasis of agglomeration as a force in shaping the economic landscape. Fujita and Thisse (2002) describe the space-economy as seen through neoclassical spectacles as a tending to a system of 'back-yard capitalism'.

The originality and value of the Krugman model as an approach to spatial analysis is its formulation of the problem in terms of monopolistic competition and increasing returns within the firm. The notion that agglomeration has its roots at least partially in monopolistic competition is particularly interesting, and corresponds well with the character of much industry today. Modern sectors such as high-technology manufacturing, business and financial services, cultural products, and so on, are especially prone to form distinctive clusters, and it is exactly in such sectors that we find the high levels of product variety, intra-sectoral trade, and the drive to market extension that characterize monopolistic competition. At the same time, the core model breathes new life into the notion of pecuniary externalities as originally formulated by Scitovsky (1954). The emphasis on agglomeration as an outcome of the complex pecuniary effects of Chamberlinian competition and internal economies of scale is unquestionably the model's principal claim to theoretical significance, and it is all the more interesting because it sets these within a framework of multi-region interdependencies.

Once all of this has been said, many reservations remain. At the outset let me state that I do not share the inclination of some geographers to discard the new geographical economics simply on the basis of its commitment to a priori forms of deductive theorizing. In practice, of course, such theorizing sometimes turns in upon itself in highly dysfunctional ways, and economists are notorious for their cultivation of an ingrown professional culture focused on displays of bravura but vacuous analytics (cf. McClosky 2002). The Krugman model and its derivative expressions certainly suffer from this syndrome, especially in view of the implausibility and arbitrariness of many of its assumptions, where enormous compromises with reality are made so as to ensure that numerical solutions can be generated,[3] and it is tempting to reject the model out of hand on the grounds of its unrealistic assumptions alone. However, I think it better to issue the main challenge from the basis of a related but slightly different perspective. In other words, what is this a model of? It may well be a description of life on some planet somewhere in the universe, but what exactly is its relevance to an understanding of economic realities on planet earth at any time in the past or the foreseeable future?

This question is underlined by the fact that the core model puts the emphasis on market-driven pecuniary relationships in an equilibrium Chamberlinian framework. In so far as it goes, this point of departure has the merit of making inter-regional competitive forces an explicit element of the analysis (as befits its intellectual origins in international trade theory). Its principal deficiency is that it fails adequately to grasp at the notion of the region as a nexus of production relationships and associated social infrastructures from which streams of external economies of scale and scope continually flow, even in single-region economies, and even in cases where competition in final product markets is non-monopolistic.

Equally, the model diverts attention away from the fact that for any kind of regional development to occur, productive assets need to be physically mobilized and integrated with one another on the ground in specific regions (Hirschman 1958). In fact, the model, as such, has virtually nothing to say about the *endogenous* intra-regional organization and dynamics of production, and almost as little about the region as a motor (as opposed to a receptacle) of economic activity (cf. Scott 2002b). In more specific terms, and despite the fact that Krugman and his co-workers make frequent reference to Marshall, the model actually gives short shrift to any meaningfully Marshallian approach to regional development and agglomeration.[4]

Four specific lacunae of the core model merit further attention in this connection:

First, the model identifies productive activity only in terms of monopolistically-competitive firms with fixed and variable costs. In its initial formulation it makes no reference whatever to the dynamics of the social division of labor and the networks of transactional relations that flow from this process. In later formulations (e.g. Krugman and Venables 1996; Venables 1996) an intermediate goods industry is assumed by fiat to exist in the model. However, the model is silent on the endogenous relations that exist in reality between the vertical structure of production and spatially dependent transactions costs. These relations tend to be of special interest and importance in clustered economic systems where intra- and inter-firm transactional structures are usually extremely complex (e.g. Scott 1983). Accordingly, the model pays inadequate attention to the wider logic of locational convergence/divergence, and, in particular, it is deficient in its grasp of the individual regional economy as a source of competitive advantage (cf. Porter 2001).

Second, these failings are compounded by the model's neglect of local labor market processes, such as information flows, job search patterns, labor-force training, and so on (Peck 1996). True enough, Krugman pays lip service to the existence of processes like these, but makes no effort to incorporate them into the workings of the core model.

Third, region-based learning and innovation processes are conspicuous by their absence from the core model. A consequence of this absence is that the core model pays little or no attention to patterns of temporal change in the qualitative attributes and competitive advantages of regional production systems. The rich parallel literature by economists such as Jaffe et al. (1993), Audretsch and Feldman (1996), or Acs (2002) on regional innovation systems compensates in some degree for this omission, but the model itself remains more or less impervious to conceptions of technology-led growth (Acs and Varga 2002).

Fourth, given its resolute commitment to microeconomic forms of analysis, the model actively suppresses the possibility that collective region-based strategies of economic adjustment might play a role in the construction of localized competitive advantages (Neary 2001). In practice, such strategies are often highly developed in regions with active production systems, both in the private sphere (e.g. inter-firm collaboration), and in the public sphere (e.g. local economic development and training programs under the aegis of regional agencies). Numerous researchers have shown time and again that strategies like these are critical to the creation of regional competitive advantages and an important tool in the search for improved rates of local economic growth (Bianchi 1992; Cooke 1999; Saxenian 1994; Storper and Scott 1995).

Some of the lacunae pointed out here can no doubt be dealt with in part by appropriate reformulations of the model (such as the introduction of commuting costs to reflect the spatial organization of local labor markets, or explicit reference to coalition formation processes), but at the cost of enormous increases of algebraic complexity. The Krugman model is for the most part a black box that occludes what by many accounts must be seen as some of the most important aspects of regional economic growth and development. As such, it casts only a very limited light on the full play of externalities, competitive advantage, and locational agglomeration in economic geography. Needless to say, the model is silent on wider social and political issues of relevance to the analysis of agglomeration, such as, for example, region-specific forms of worker socialization and habituation, the emergence of local governance structures, or the historical shifts that occur periodically in technical-organizational structures of accumulation, and that greatly impact regional trajectories of development.

By its elevation of atomistic exchange relations to an exclusive ontology of the economic and the geographic (albeit in a Chamberlinian context), the core model provides only a very partial account of the genesis and logic of the economic landscape. In the end, the model can be seen more as an effort to codify atomized market processes in simple spatial frameworks than it is an attempt to understand spatial relations in any thorough-going sense of the term. The strong point of the model is its description of pecuniary externalities in multi-region systems; the weak point is its account of locational adjustment in which units of capital and labor move like billiard balls (except for the ones that have been nailed to the table) from one equilibrium to another, and then simply fall magically into a fully functional economic system as they accumulate in receiving regions. Nevertheless – and to be fair to the wider body of urban and regional economics generally – there is an obvious and encouraging trend in much of the current literature to move beyond the limitations of the core model as expounded by Krugman and to deal in a more flexible and open-ended manner with many of the issues where it is most vulnerable to criticism (see, for example, the essays collected together in Cheshire and Mills 1999; and Henderson and Thisse 2004).

Capitalism, culture, and geography

Economic geographers discover culture

Just as geographical economics was making its appearance on the academic scene in the early 1990s, a number of economic geographers were transferring their attentions to an altogether different set of approaches rooted in issues of culture. The emergence of this interest coincided with a growing conviction that not only were certain earlier generations of geographers and other social scientists incorrect to regard culture simply as an outcome of underlying economic realities, but that these realities themselves are in fundamental ways subject to the play of cultural forces.

Any casual scrutiny of contemporary capitalism reveals at once that it is inflected with different social and cultural resonances in different localities, and that these resonances are directly implicated in the organization of economic life and modalities of economic calculation. American, Japanese, and Chinese capitalism, for example, are at once generically similar and yet are marked by socio-cultural idiosyncrasies with significant effects on the ways in which they function. In addition, production systems in contemporary capitalism, while still obviously highly focused on the mechanical manufacture of things, are shifting more and more into the processing of information and symbols, from business advice to cultural services. This trend is leading to dramatic changes in the form and function of commodified goods and services, and much research is now moving forward on the reception, interpretation, and social effects of these outputs (Bridge and Smith 2003; Jackson 1999; Thrift 2000). This research points to important theoretical issues concerning the hermeneutics of the commodity, and the functions of capitalism generally as a fountainhead of symbolic representation in modern life (e.g. Harvey 1989; Lash and Urry 1994). Striking changes are also occurring in the social and psychological make-up of the workplace. Over large areas of the new economy of capitalism, dress, mannerisms, forms of speech, self-presentation, and so on have become essential elements of workers' performance. Equally, gender, race, ethnicity, and so on, together with specific forms of empathy associated with them are being actively exploited and managed in various ways in the workplace (cf. McDowell 1997). More generally, markets as a whole could not work in the absence of a sociocultural system regulating the conventions and behaviors that sustain them.

These brief remarks, schematic as they may be, already underline the obvious and pressing need for economic geographers to pay close attention to the ways in which culture and economy intersect with one another in mutually constitutive ways. The urgency of this need is reinforced by the observation that the economic and the cultural come together with special intensity in place (Shields 1999), and that many of the key agglomerations constituting the focal points of the new economy around the world are critically dependent on the complex play of culture. Thus, to an ever-increasing degree, the productive performance of agglomerations like the City of London (Thrift 1994), Hollywood (Scott 2002a), or Silicon Valley (Saxenian 1994) can only be understood in relation to their joint economic and cultural dynamics. Each of these places is shot through with distinctive traditions, sensibilities, and cultural practices that leave deep imprints on phenomena such as management styles, norms of worker habituation, creative and innovative energies, the design of final outputs, and so on, and these phenomena in turn are strongly implicated in processes of local economic growth and development.

For culture; against the cultural turn

In view of the discussion above, it seems fairly safe to say that only a few die-hards and philistines are likely to make strenuous objections to attempts to bring culture more forcefully into the study of economic geography. In spite of the neologisms and cliché-ridden prose that Martin and Sunley (2001) rightly complain about, there is obviously a significant nexus of ideas in a more culturally-inflected economic geography that responds in a very genuine way to major problems posed by contemporary capitalist society. Once this point has been made, however, a number of the reforms of economic geography that have been most strenuously advocated under the rubric of the cultural turn are rather less obviously acceptable, and have recently been subject to heated debate by economic geographers (see, for example, Martin and Sunley 2001; Plummer and Sheppard 2001; Rodriguez-Pose 2001; Sayer 1997; Storper 2001).

This debate has tended to find its sharpest expression in relation to the curious reluctance by some proponents of the cultural turn to make any concession to the play of economic processes in economic geography except in so far as they are an expression of underlying cultural dynamics. In a number of their more fervent statements, indeed, some of these proponents occasionally verge on an inversion of the classical Marxian conceit to the effect that culture flows uni-causally from the economy, by offering equally exaggerated claims about the influence of culture on the economy. In a statement that displays much enthusiasm about the study of culture and much acrimony in regard to the discipline of economics, Amin and Thrift (2000) essentially recommend withdrawal from economic analysis, as such, and a wholesale re-description of economic realities in terms of cultural points of reference. Thus, in writing about the problem of eventuation they one-sidedly argue that 'acting into the words confirms the discourse and makes a new real' (p. 6), so that in their formulation, the economy becomes nothing more than a series of 'performances' derived from a script. Elsewhere, Thrift (2001) further proclaims that the new economy of the 1990s was fundamentally a rhetorical phenomenon. The argument here starts off promisingly enough with an examination of the role of the press, business consultants, financial advisors, and the like, in helping to foment the fast-paced, high-risk economic environment of the period, but then it veers into the blunt assertion that the new economy as a whole can be understood simply as a discursive construct. In formulations like these, basic economic realities – the state of technology, the rhythms of capital accumulation and investment, the rate of profit, the flow of circulating capital, and so on – become just so much inert plasma to be written upon this way or that as cultural shifts occur and as revisions of the script are introduced. Certainly, words are a critical moment in the circuit of mediations through which economic reality operates, and there can be no doubt that many unique effects are set in motion at this particular level of analysis. Conversely, and it is puzzling that so trivial and obvious a point should need to be made, there are also deeply-rooted economic logics and dynamics at work in the contemporary space-economy, and at least some of these (such as the dynamics of industrial organization, or the increasing returns effects that lie at the root of industrial districts), require investigation on their own terms above and beyond invocations of the causal powers of discourse and culture.

In a series of recent writings, Barnes (e.g. 1996, 2001, 2003), has pursued a related line of investigation opened up by the cultural turn. Barnes' work is much influenced by Derrida and Rorty, and is centrally focused on the metaphorical and narratological character of geographical writing. There is actually much of interest in the approach Barnes takes. He has many useful things to say about the ideologies and working habits of economic geographers, as well as about the rhetorical devices that they deploy in their written reports. This helps among other things to keep us focused on the critical idea that our intellectual encounters with the real are always deeply theory-dependent (Sunley 1996). But as the plot thickens – or thins, according to your taste – we steadily lose sight of economic geography as a discipline with concrete substantive concerns (such as regional development or income inequalities), for these simply dissolve away into the primacy of the text and its metaphorical perplexities. I am perfectly prepared to admit that there may be strong elements of metaphor in, for example, a geography of hunger, but I certainly have no sympathy for the idea that hunger is just a metaphor, if only on the ad hominem grounds that it has painful physical manifestations and morbid long-term effects. Here, the legitimate claim that we can only know the world through socially-constructed codes of reference seems to have given way to the sophism that all we can know about the world is the codes themselves.

An even more extreme case of the solipsism that haunts much of the cultural turn can be found in the book by Gibson-Graham (1996) about strategic possibilities for progressive social change in contemporary capitalism. The central arguments of the book hinge upon the proposition that the criteria for validating a theory are purely internal to the theory to be validated. As Gibson-Graham writes (p. 60): 'We cannot argue that our theory has more explanatory power or greater proximity to the truth than other theories because there is no common standard which could serve as the instrument of such a metatheoretical validation process'. If this proposition were indeed true it would presumably undermine much of the point in Gibson-Graham proceeding any further in her argument, though she does in fact continue on for another 200-odd pages. In the course of this discussion, the relativism of her main thesis is steadily transformed from a merely academic exercise into a political agenda of sorts. Thus, she announces (p. 260), 'the way to begin to break free of capitalism is to turn its prevalent representations on their heads'. Presto. Not even a hint about a possible transitional program, or a few suggestions about, say, practical reform of the banking system. The claim is presented in all its baldness, without any apparent consciousness that attempts to break free of any given social system are likely to run into the stubborn realities of its indurated social and property relations as they actually exist. More generally, Gibson-Graham's argument leads inexorably beyond the perfectly acceptable notion that all intellectual work is theory-dependent and into those murky tracts of idealist philosophy where reality is merely a reflection of theory, and where theory produces social change independently of concrete practice and disciplined attention to the refractory resistances of things as they really are.

So, quite apart from its dysfunctional depreciation of the role of economic forces and structural logics in economic geography, the cultural turn also opens a door to a disconcerting strain of philosophical idealism and political voluntarism in modern geography. The net effect is what we might call economistic grand theory in reverse: a remarkable failure to recognize sensible boundaries as to just what precisely a cultural theory of the economy can achieve, and a concomitant over-promotion of the notion that social and economic transformation involves nothing more than the unmediated power of theoretical ideas. Again, nothing in this argument is intended to deny the important continuities and intersections between culture and economy or the significance of the economy as a site of cultural practices; neither is it in any sense an attempt to eject the study of cultural economy from geography. The problem is not 'culture' but the cultural turn as it has emerged out of cultural studies with its militant project of reinterpreting all social relations as cultural relations, and its naïve, if understandable, attempt to humanize the iron cage of capitalist accumulation by unwarranted culturalization of its central economic dynamics (Eagleton 2003; Rojek and Turner 2000).

Toward a re-synthesis

As I have tried to show in this chapter, Krugman-style geographical economics offers at best an extremely narrow vision of the dynamics of the economic landscape, and is in any case, less preoccupied with geography as such than it is with geography as just another domain within which markets unfold. Geographers and economists certainly occupy much common ground at the present time, but encounters between them on this shared terrain are endemically susceptible to deeply-seated disputes about theoretical priorities. My guess is that the influence of the Krugman model will in any case soon wither away as geographical economics comes up against the model's inner and outer limits, just as neoclassical regional science began to show signs of enervation after the mid-1970s in part as a consequence of its commitment to the strait-jacket of convexity and constant returns to scale (cf. Neary 2001; Thisse 1997). As it happens, any such retreat may just possibly help to open up opportunities for more fruitful future encounters between geographers and economists over issues of space (see also Sjöberg and Sjöholm 2002). For the present, the undisputed major contribution of geographical economics to our understanding of spatial problems has been its resuscitation of the notion of pecuniary externalities in a world of Chamberlinian competition.

The cultural turn, for its part, has sought to take economic geography in an altogether different direction. In some degree, of course, the clashing claims of economic geographers and cultural geographers over the last decade or so can be interpreted as expressions of an internal power struggle for status and influence in the profession of geography as a whole. This struggle owes much to the unquestioned intellectual re-invigoration and consequent self-assertion of cultural geography that occurred over the 1990s as cultural studies expanded in the academy at large. Despite the clashes, there remains, as I have indicated, much useful work to be accomplished by cooperation between economic and cultural geographers in any effort to comprehend the spatiality and locational dynamics of modern capitalism (Bathelt and Glückler 2003; Gertler 2003a; Gregson et al. 2001; Yeung 2003). At the same time, there will undoubtedly continue to be strong points of divergence between the two subdisciplines; lines of investigation opened up by economic geographers where cultural geographers hesitate to tread, and vice versa. A degree of mutual tolerance (though certainly not automatic and uncritical mutual endorsement) is no doubt called for in this situation.

Notwithstanding all the theoretical turbulence of the last few decades, there is probably still wide agreement among economic geographers, as such, that one of the main tasks we face is in the end some sort of transformative understanding of the historical geography of capitalist society (Harvey 1982; Harvey and Scott 1989). I suspect, as well, that most economic geographers would agree with the proposition that we need some sort of new synthesis in order to pursue this task more effectively (cf. Castree 1999), i.e. a revised cognitive map that can help us make sense of all the complex contemporary tendencies that have turned what critical theorists used hopefully to call 'late capitalism' into the triumphant and rejuvenated juggernaut that it is today. I make this claim about the need for a new synthesis in full cognizance of the reductionist dangers that it opens up (cf. Amin and Robbins 1990; Sayer 2000). Equally, I want to avoid the self-defeating conclusion that because of these dangers we must always downsize our theoretical ambitions. One of the truly disconcerting aspects of much geographical work today is that it preaches a doctrine that privileges the small, the piecemeal, and the local, even as capital plays out its own grandiose saga of expansion and recuperation at an increasingly globalized scale.

A prospective economic geography capable of dealing with the contemporary world must hew closely, it seems to me, to the following programmatic goals if it is to achieve a powerful purchase on both scientific insight and progressive political strategy.

- To begin at the beginning: economic geography needs to work out a theoretical re-description of capitalism as a structure of production and consumption and as an engine of accumulation, taking into account the dramatic changes that have occurred in recent decades in such phenomena as technology, forms of industrial and corporate organization, financial systems, labor markets, and so on. This theoretical re-description must be sensitive to the generic or quasi-generic forms of capitalist development that occur in different times in different places, which, in turn, entails attention to the kinds of issues that regulation theorists have identified under the general rubric of regimes of accumulation (Aglietta 1976; Lipietz 1986).

- In addition to these economic concerns, we must recognize that contemporary capitalism is intertwined with enormously heterogeneous forms of social and cultural life, and that no one element of this conjoint field is necessarily reducible to the other. Directions of causality and influence across this field are a matter of empirical investigation, not of theoretical pre-judgment. Note that in this formulation, class becomes only one possible dimension of social existence out of a multiplicity of other actual and possible dimensions.
- This nexus of economic, social, and cultural relationships constitutes a creative field or environment within which complex processes of entrepreneurship, learning, and innovation occur. Geographers have a special interest in deciphering the spatial logic of this field and in demonstrating how it helps to shape locational dynamics.
- In combination with these modalities of economic and social reality, we need to reserve a specific analytical and descriptive space for collective action and institutional order at many different levels of spatial and organizational scale (the firm, the local labor market, the region, the nation, etc.), together with a due sense of the political tensions and rivalries that run throughout this sphere of human development. By the same token, a vibrant economic geography will always not only be openly policy-relevant (Markusen 1999), but also politically engaged. A key question in this context is how to build local institutional frameworks that promote both economic success and social justice.
- We must recognize that social and economic relations are often extremely durable, and that they have a propensity to become independent in varying degree of the individuals caught up within them. This means that any normative account of social transformation and political strategy, must deal seriously with the idea that there are likely to be stubborn resistances to change rooted in these same relations. The solutions to this problem proposed by sociologists like Bourdieu (1972) and Giddens (1979) strike me as providing reasonable bases for pushing forward in this respect, for they explicitly recognize the inertia of social structures while simultaneously insisting on the integrity of individual human volition. Unfortunately, these solutions (most especially the structure-agency formulation of Giddens) have been much diluted in recent years by reinterpretations that lean increasingly heavily on the agency side of the equation, partly as a reflection of the cultural turn, partly out of a misplaced fear of falling into the pit of determinism.[5] Invocations of unmediated agency (or, for that matter, neoclassical utility) as an explanatory variable in social science are often little more than confessions of ignorance, in the sense that when we are unable to account for certain kinds of relationships or events, we are often tempted to fall back on the reassuring notion that things are thus and so for no other reason than because that's the way we want them to be, irrespective of any underlying structural conditions.
- A corollary of the structured organization and sunk costs of social life is that economic relationships (especially when they are locationally interrelated,

as in the case of a regional production system) are likely to be path-dependent. This observation suggests at once that an evolutionary perspective is well suited to capture important elements of the dynamics of the economic landscape (cf. Boschma and Lambooy 1999; Nelson and Winter 1982). It follows that any attempt to describe the economic landscape in terms of instantaneous adjustment and readjustment to a neoclassical optimum optimorum is intrinsically irrelevant.

- All of these moments of economic and social reality occur in a world in which geography has not yet been—and cannot yet be—abolished (Leamer and Storper 2001). The dynamics of accumulation shape geographic space, and equally importantly, geographic space shapes the dynamics of accumulation. This means, too, that capitalism is differentiated at varying levels of spatial resolution, from the local to the global, and that sharp differences occur in forms of life from place to place. Indeed, as globalization now begins to run its course, geographic space becomes more important, not less important, because it presents ever-widening possibilities for finely-grained locational specialization and differentiation. Critical analysis of these possibilities must be one of modern economic geography's principal concerns.
- Finally, I want to enter a plea for methodological variety and openness. One corollary of this plea is that economic geographers need to recover the lost skills of quantitative analysis, not out of some atavistic impulse to reinstate the economic geography of the 1960s, but because of the proven value of these skills in the investigation of economic data. The steady erosion of geographers' capabilities in this regard over the last couple of decades is surely a net loss to the discipline.

These remarks still leave open a wide range of alternative research strategies and theoretical orientations in economic geography, including approaches marked variously by heavy doses of algebraic formalization or cultural commentary. A particular point of focus, however, is provided by the continuing commitment by significant numbers of economic geographers to critical analysis and to the search for progressive social change. The pursuit of some sort of social democratic agenda and the fight against global neoliberalism, it seems to me, must stand high in any set of priorities in this regard at the present time, and all the more so as our world remains an arena in which tremendous variations in living standards, economic opportunity, and possibilities for cultural self-realization persist tenaciously from place to place and country to country. More than anything else, the great testing ground for economic geography, now and in the foreseeable future, must surely be identified in one way or another in relation to the central question of development, not only in its expression as a problem in historical geography, but as a normative project of global significance.

Economic geographers have much work to do in dealing with the multiple challenges of this evolving situation. But they also need periodically to take critical soundings of their tools, their practices, and their theoretical commitments if they are to remain equal to the daunting tasks ahead.

Acknowledgements

I am grateful to Jeffrey Boggs, Steven Brakman, Harry Garretsen, David Rigby, Michael Storper, and Jacques Thisse for their comments on an earlier draft of this chapter. None of these individuals bears any responsibility for the opinions expressed here.

Notes

1. Because of the model's complexity, it is typically defined in terms of just two regions for expository purposes.
2. Other work in contemporary geographical economics (not all of it strictly in line with the core model) include regional income inequalities (Barro and Sala-i-Martin 1995; Quah 1996), the dynamics of city systems (Duranton and Puga 2000; Ellison and Glaeser 1997), regional productivity and growth (Henderson 2003), and so on. My remarks in this chapter are focused on the narrower (Krugmanian) view of geographical economics, both because of the extremely large claims that have been made on its behalf and because of its current centrality in the entire project of geographical economics (but see my later assessment of how this situation may change).
3. Some sectors are taken to be monopolistically competitive; others are deemed subject to perfect competition where this is analytically convenient. Elasticities of demand and substitution are always held constant. Firms have no opportunities for strategic interaction. The multidimensional character of business transactions is reduced to the fiction of iceberg transport costs. Labor is mobile when the algebra demands it; labor is immobile otherwise. Wages are adjustable in some cases, sticky in others. Generalizations of the model to more than two regions result in a world that is shaped like a doughnut, not because this makes any sense in substantive terms but because the mathematics are otherwise intractable. The model contains fixed costs but no sunk costs, and there are thus no inertial barriers to adjustment in the model (where adjustment proceeds relatively rapidly whereas adjustment of the economic landscape in reality tends to be extremely slow). And what exactly is the model's appropriate spatial scale of resolution? Ottaviano and Thisse (2001) appear to feel, with some justification I believe, that the model of pecuniary externalities works best at a level of resolution where regions approximate the size of the US Manufacturing Belt. The model certainly does not seem to have much relevance to cases where regions are defined at small scales of spatial resolution. For a more extended discussion of the problem of scale in the new geographical economics, see Olsen (2002).
4. Sheppard (2000) makes much the same point. For examples of the Marshallian approach as developed by geographers see Amin and Thrift (1992), Cooke and Morgan (1998), Gertler (2003b), Rigby and Essletzbichler (2002), Scott (1988), Storper (1997).
5. It is useful here to recall the early argument of Martin (1951) to the effect that any self-respecting determinism insists on a direct mechanistic link from matter or the external world to mind so that what passes for free will is (so the determinist would say) nothing more than a cause-effect relationship. The existence of structural constraints on human action, or even the emergence of common social predispositions and habits, do not, by this standard of judgment, amount to any form of determinism. Nor can determinism in this rigorous sense necessarily be equated with the existence of macro-social outcomes that occur independently of any explicit decision that the world should be structured thus and so, or with situations where these outcomes assume 'laws of motion' without our explicit permission, as it were (e.g. the pervasive separation of home and work in the modern metropolis and the daily waves of commuting that are a result of this circumstance). Mutatis mutandis, when geographers invoke

unmediated 'agency' or 'volition' as an explanatory variable, they are implicitly confessing to a failure of analysis, even though agency and volition are always a component of any human action.

References

Abdel-Rahman, H. and M. Fujita (1990) 'Product variety, Marshallian externalities, and city sizes'. *Journal of Regional Science,* 30: 165–83.

Acs, Z. J. (2002) *Innovation and the Growth of Cities.* Cheltenham: Edward Elgar.

Acs, Z. J. and A. Varga (2002) 'Geography, endogenous growth and innovation'. *International Regional Science Review,* 25: 132–48.

Aglietta, M. (1976) *Régulation et Crises du Capitalisme.* Paris: Calmann-Lévy.

Amin, A. and K. Robbins (1990) 'The re-emergence of regional economies? The mythical geography of flexible accumulation'. *Environment and Planning D: Society and Space,* 8: 7–34.

Amin, A. and N. J. Thrift (2000) 'What kind of economic theory for what kind of economic geography?' *Antipode,* 32: 4–9.

Amin, A. and N. J. Thrift (1992) 'Neo-Marshallian nodes in global networks'. *International Journal of Urban and Regional Research,* 16: 571–81.

Audretsch, D. R. and M. P. Feldman (1996) 'R&D spillovers and the geography of innovation and production'. *American Economic Review,* 86: 630–40.

Baldwin, R. E. (1999) 'Agglomeration and endogenous capital'. *European Economic Review,* 43: 253–80.

Barnes, B. (1974) *Scientific Knowledge and Sociological Theory.* London: Routledge and Kegan Paul.

Barnes, T. J. (1996) *Logics of Dislocation: Models, Metaphors, and Meanings of Economic Space.* New York: Guilford.

Barnes, T. J. (2001) 'Retheorizing geography: from the quantitative revolution to the cultural turn'. *Annals of the Association of American Geographers,* 91: 546–65.

Barnes, T. J. (2003) 'The place of locational analysis: a selective and interpretive history'. *Progress in Human Geography,* 27: 69–95.

Barro, R. J. and X. Sala-i-Martin (1995) *Economic Growth.* New York: McGraw-Hill.

Bathelt, H. and J. Glückler (2003) 'Toward a relational economic geography'. *Journal of Economic Geography,* 3: 117–44.

Becattini, G. (1987) *Mercato e Forze Local: il Distretto Industriale.* Bologna: Il Mulino.

Benko, G. (1998) *La Science Régionale.* Paris: Presses Universitaires de France.

Bianchi, P. (1992) 'Levels of policy and the nature of post-fordist competition'. In M. Storper and A. J. Scott (eds) *Pathways to Industrialization and Regional Development,* pp. 303–15. London: Routledge.

Bluestone, B. and B. Harrison (1982) *The Deindustrialization of America.* New York: Basic Books.

Boddy, M. (1999) 'Geographical economics and urban competitiveness: a critique'. *Urban Studies,* 36: 811–42.

Boggs, J. S. and N. M. Rantisi (2003) 'The relational turn in economic geography'. *Journal of Economic Geography,* 3: 109–16.

Boschma, R. A. and J. G. Lambooy (1999) 'Evolutionary economics and economic geography'. *Journal of Evolutionary Economics,* 9: 411–29.

Bourdieu, P. (1972) *Esquisse d'une Théorie de la Pratique.* Geneva: Librairie Droz.

Bridge, G. and A. Smith (2003) 'Intimate encounters: culture—economy—commodity'. *Environment and Planning D: Society and Space*, 21: 251–68.

Carney, J., R. Hudson and J. Lewis (1980) *Regions in Crisis: New Perspective in European Regional Theory.* New York: St Martin's Press.

Castree, N. (1999) 'Envisioning capitalism: geography and the renewal of Marxian political economy'. *Transactions of the Institute of British Geographers*, 24: 137–58.

Chamberlin, E. (1933) *The Theory of Monopolistic Competition.* Cambridge, MA: Harvard University Press.

Cheshire, P. and E. S. Mills (1999) *Handbook of Regional and Urban Economics, Volume 3, Applied Urban Economics.* Amsterdam: North Holland.

Clark, G. L. (1998) 'Stylized facts and close dialogue: methodology in economic geography'. *Annals of the Association of American Geographers*, 88: 73–87.

Cooke, P. (1999) 'The co-operative advantage of regions'. In T. J. Barnes and M. S. Gertler (eds) *The New Industrial Geography: Regions, Regulation and Institutions*, pp. 54–73. London: Routledge.

Cooke, P. and K. Morgan (1998) *The Associational Economy: Firms, Regions, and Innovation.* Oxford: Oxford University Press.

Crang, P. (1997) 'Introduction: cultural turns and the (re)constitution of economic geography'. In J. Wills and R. Lee (eds), *Geographies of Economies*, pp. 3–15. London: Arnold.

David, P. A. (1999) 'Krugman's economic geography of development: NEGs, POGs and naked models in space'. *International Regional Science Review*, 22: 162–72.

Dear, M. J. (2000) *The Postmodern Urban Condition.* Oxford: Blackwell.

Dicken, P. (1992) *Global Shift: The Internationalization of Economic Activity.* New York: Guilford.

Dixit, A. K. and J. E. Stiglitz (1977) 'Monopolistic competition and optimum product diversity'. *American Economic Review*, 67: 297–308.

Duranton, G. and D. Puga (2000) 'Diversity and specialization in cities: why, where and when does it matter?' *Urban Studies*, 37: 533–55.

Dymski, G. (1996) 'On Krugman's model of economic geography'. *Geoforum*, 27: 439–52.

Eagleton, T. (2003) *After Theory.* New York: Basic Books.

Ellison, G. and E. L. Glaeser (1997) 'Geographic concentration in US manufacturing industries: a dartboard approach'. *Journal of Political Economy*, 105: 889–927.

Fröbel, F., J. Heinrichs and O. Kreye (1980) *The New International Division of Labor.* Cambridge: Cambridge University Press.

Fujita, M. and J.-F. Thisse (2002) *Economics of Agglomeration: Cities, Industrial Location, and Regional Growth.* Cambridge: Cambridge University Press.

Geertz, C. (1983) *Local Knowledge: Further Essays in Interpretive Anthropology.* New York: Basic Books.

Gertler, M. S. (2003a) 'A cultural economic geography of production'. In K. Anderson, M. Domosh, S. Pile and N. Thrift (eds), pp. 131–46. *Handbook of Cultural Geography.* London: Sage.

Gertler, M. S. (2003b) 'Tacit knowledge and the economic geography of context, or, the undefinable tacitness of being (there)'. *Journal of Economic Geography*, 3: 75–99.

Gibson-Graham, J. K. (1996) *The End of Capitalism (As We Knew it): A Feminist Critique of Political Economy.* Oxford: Blackwell.

Giddens, A. (1979) *Central Problems in Social Theory: Action, Structure and Contradiction in Social Analysis.* London: Macmillan.

Gould, P. (1979) 'Geography 1957–1977: the Augean period'. *Annals of the Association of American Geographers*, 53: 290–7.

Gregson, N., K. Simonsen and D. Vaiou (2001) 'Whose economy for whose culture? Moving beyond oppositional talk in European debate about economy and culture'. *Antipode*, 33: 616–46.

Haraway, D. J. (1991) *Simians, Cyborgs and Women: the Reinvention of Nature*. New York: Routledge.

Harvey, D. (1982) *The Limits to Capital*. Oxford: Blackwell.

Harvey, D. (1989) *The Condition of Postmodernity: An Enquiry into the Origins of Cultural Change*. Oxford: Blackwell.

Harvey, D. and A. J. Scott (1989) 'The practice of human geography: theory and empirical specificity in the transition from fordism to flexible accumulation'. In B. Macmillan (ed.), pp. 217–29. *Remodelling Geography*. Oxford: Blackwell.

Henderson, V. (2003) 'The urbanization process and economic growth: the so-what question'. *Journal of Economic Growth*, 8: 47–71.

Henderson, V. and J. F. Thisse (2004) *Handbook of Regional and Urban Economics, Volume 4, Cities and Geography*. Amsterdam: North-Holland.

Hirschman, A. O. (1958) *The Strategy of Economic Development*. New Haven, CT: Yale University Press.

Imrie, R., S. Pinch and M. Boyle (1996) 'Identities, citizenship and power in the cities'. *Urban Studies*, 33: 1255–62.

Jackson, P. (1999) 'Commodity culture: the traffic in things'. *Transactions of the Institute of British Geographers*, 24: 95–108.

Jaffe, A. B., M. Trajtenberg and R. Henderson (1993) 'Geographic localization of knowledge spillovers as evidenced by patent citations'. *Quarterly Journal of Economics*, 108: 577–98.

Johnston, R. J. (1992) 'Review of geography and trade'. *Environment and Planning A*, 24: 1066.

Johnston, R. J., P. J. Taylor and M. J. Watts (1995) *Geographies of Global Change*. Oxford: Blackwell.

Kaldor, N. (1970) 'The case for regional policies'. *Scottish Journal of Political Economy*, 17: 337–48.

Krugman, P. (1991) *Geography and Trade*. Leuven: Leuven University Press.

Krugman, P. and A. J. Venables (1995) 'Globalization and the inequality of nations'. *Quarterly Journal of Economics*, 110: 857–80.

Krugman, P. and A. J. Venables (1996) 'Integration, specialization, and adjustment'. *European Economic Review*, 40: 959–67.

Lash, S. and J. Urry (1994) *Economies of Signs and Space*. London, Thousand Oaks: Sage.

Latour, B. (1991) *Nous n'avons jamais été modernes: essai d'anthropologie symétrique*. Paris: La Découverte.

Leamer, E. E. and M. Storper (2001) 'The economic geography of the Internet age'. *Journal of International Business Studies*, 32: 641–65.

Lee, R. (2002) 'Nice maps, shame about the theory: thinking geographically about the economic'. *Progress in Human Geography*, 26: 333–55.

Lipietz, A. (1986) 'New tendencies in the international division of labor: regimes of accumulation and modes of social regulation'. In A. J. Scott and M. Storper (eds), pp. 16–40. *Production, Work, Territory: The Anatomy of Industrial Capitalism*. Boston: Allen and Unwin.

Mannheim, K. (1952) *Essays in the Sociology of Knowledge*. Henley-on-Thames: Routledge and Kegan Paul.

Markusen, A. (1999) 'Fuzzy concepts, scanty evidence, policy distance: the case for rigour and policy relevance in critical regional studies'. *Regional Studies,* 33: 869–84.

Martin, A. F. (1951) 'The necessity for determinism'. *Transactions of the Institute of British Geographers,* 17: 1–11.

Martin, R. (1999) 'The new geographical turn in economics: some reflections'. *Cambridge Journal of Economics,* 23: 65–91.

Martin, R. (2001) 'Geography and public policy: the case of the missing agenda'. *Progress in Human Geography,* 25: 189–210.

Martin, R. and P. Sunley (2001) 'Rethinking the 'economic' in economic geography: broadening our vision or losing our focus?' *Antipode,* 33: 148–61.

Massey, D. and R. Meegan (1982) *Anatomy of Job Loss: The How, Why, Where, and When of Employment Decline.* London: Methuen.

McClosky, D. (2002) *The Secret Sins of Economists.* Chicago: Prickly Paradigm Press.

McDowell, L. (1997) *Capital Culture: Gender at Work in the City.* Oxford: Blackwell.

Meardon, S. J. (2000) 'Eclecticism, inconsistency, and innovation in the history of geographical economics'. In R. E. Backhouse and J. Biddle (eds), pp. 325–59. *Toward a History of Applied Economics.* Durham, NC: Duke University Press.

Myrdal, G. (1959) *Economic Theory and Under-Developed Regions.* London: Gerald Duckworth.

Neary, J. P. (2001) 'Of hype and hyperbolas: introducing the new economic geography'. *Journal of Economic Literature,* 39: 536–61.

Nelson, R. R. and S. G. Winter (1982) *An Evolutionary Theory of Economic Change.* Cambridge, MA: Belknap Press.

Olsen, J. (2002) 'On the units of geographical economics'. *Geoforum,* 33: 153–64.

Ottaviano, G. I. P. and J. F. Thisse (2001) 'On economic geography in economic theory: increasing returns and pecuniary externalities'. *Journal of Economic Geography,* 1: 153–79.

Peck, J. (1996) *Work-Place: the Social Regulation of Labor Markets.* New York: Guilford Press.

Plummer, P. and E. Sheppard (2001) 'Must emancipatory economic geography be qualitative?' *Antipode,* 33: 194–9.

Porter, M. E. (2001) 'Regions and the new economics of competition'. In A. J. Scott (ed.), pp. 139–57. *Global City-Regions: Trends, Theory, Policy.* Oxford: Oxford University Press.

Quah, D. T. (1996) 'Regional convergence clusters across Europe'. *European Economic Review,* 40: 951–8.

Rigby, D. L. and J. Essletzbichler (2002) 'Agglomeration economies and productivity differences in US cities'. *Journal of Economic Geography,* 2: 407–32.

Rodriguez-Pose, A. (2001) 'Killing economic geography with a 'cultural turn' overdose'. *Antipode,* 33: 176–82.

Rojek, C. and B. Turner (2000) 'Decorative sociology: towards a critique of the cultural turn'. *The Sociological Review,* 48: 629–48.

Rorty, R. (1979) *Philosophy and the Mirror of Nature.* Princeton, NJ: Princeton University Press.

Saxenian, A. (1994) *Regional Advantage: Culture and Competition in Silicon Valley and Route 128.* Cambridge, MA: Harvard University Press.

Sayer, A. (1997) 'The dialectic of culture and economy'. In R. Lee and J. Wills (eds), pp. 16–26. *Geographies of Economies.* London: Arnold.

Sayer, A. (2000) *Realism and Social Science*. London: Sage.

Sayer, A. and M. Storper (1997) 'Ethics unbound: for a normative turn in social theory'. *Environment and Planning D: Society and Space*, 15: 1–17.

Scitovsky, T. (1954) 'Two concepts of external economies'. *Journal of Political Economy*, 62: 143–51.

Scott, A. J. (1983) 'Industrial organization and the logic of intra-metropolitan location: I. Theoretical considerations'. *Economic Geography*, 59: 233–50.

Scott, A. J. (1986) 'High technology industry and territorial development: the rise of the Orange County complex, 1955–1984'. *Urban Geography*, 7: 3–45.

Scott, A. J. (1988) *New Industrial Spaces: Flexible Production Organization and Regional Development In North America and Western Europe*. London: Pion.

Scott, A. J. (2000) 'Economic geography: the great half-century'. *Cambridge Journal of Economics*, 24: 483–504.

Scott, A. J. (2002a) 'A new map of Hollywood: the production and distribution of American motion pictures'. *Regional Studies*, 36: 957–75.

Scott, A. J. (2002b) 'Regional push: the geography of development and growth in low- and middle-income countries'. *Third World Quarterly*, 23: 137–61.

Shapin, S. (1998) 'Placing the view from nowhere: historical and sociological problems in the location of science'. *Transactions of the Institute of British Geographers*, 23: 5–12.

Sheppard, E. (2000) 'Geography or economics? Conceptions of space, time, interdependence, and agency'. In G. L. Clark, M. P. Feldman and M. S. Gertler (eds), pp. 99–119. *The Oxford Handbook of Economic Geography*. Oxford: Oxford University Press.

Shields, R. (1999) 'Culture and the economy of cities'. *European Urban and Regional Studies*, 6: 303–11.

Sjöberg, O. and F. Sjöholm (2002) 'Common ground? Prospects for integrating the economic geography of geographers and economists'. *Environment and Planning A*, 34: 467–86.

Smith, N. (1987) 'Dangers of the empirical turn: some comments on the CURS initiative'. *Antipode*, 19: 59–68.

Storper, M. (1997) *The Regional World: Territorial Development in a Global Economy*. New York: Guilford Press.

Storper, M. (2001) 'The poverty of radical theory today: from the false promises of Marxism to the mirage of the cultural turn'. *International Journal of Urban and Regional Research*, 25: 155–79.

Storper, M. and A. J. Scott (1995) 'The wealth of regions: market forces and policy imperatives in local and global context'. *Futures*, 27: 505–26.

Sunley, P. (1996) 'Context in economic geography: the relevance of pragmatism'. *Progress in Human Geography*, 20: 338–55.

Swyngedouw, E. (1997) 'Neither global nor local: 'globalization' and politics of scale'. In K. R. Cox (ed.), pp. 137–66. *Spaces of Globalization: Reasserting the Power of the Local*. New York: Guilford.

Taylor, P. J., G. Catalano and D. R. F. Walker (2002) 'Measurement of the world city network'. *Urban Studies*, 39: 2367–76.

Thisse, J. F. (1997) 'L'oubli de l'espace dans la pensée économique'. *Région et Développement*, 13–39.

Thrift, N. (1994) 'On the social and cultural determinants of international financial centres'. In S. Corbridge, N. Thrift and R. Martin (eds), pp. 327–55. *Money, Power and Space*. Oxford: Blackwell.

Thrift, N. (2000) 'Pandora's box? cultural geographies of economies'. In G. L. Clark, M. P. Feldman and M. S. Gertle (eds), pp. 689–704. *The Oxford Handbook of Economic Geography*. Oxford: Oxford University Press.

Thrift, N. (2001) 'It's the romance, not the finance, that makes the business worth pursuing': disclosing a new market culture'. *Economy and Society*, 30: 412–32.

Thrift, N. and K. Olds (1996) 'Refiguring the economic in economic geography'. *Progress in Human Geography*, 20: 311–37.

Venables, A. J. (1996) 'Equilibrium locations of vertically linked industries'. *International Economic Review*, 37: 341–59.

Yeung, H. W. (2003) 'Practicing new economic geographies: a methodological examination'. *Annals of the Association of American Geographers*, 93: 445–66.

Section II

Globalization and contemporary capitalism

6 Setting the agenda

The geography of global finance

Gordon L. Clark

Introduction

Economic geography has a long history, with its own historians of thought (Scott 2000, 2004). At the same time, it is a field constantly evolving seeking to incorporate the realities of modern capitalism into the scope of its responsibilities (Dicken 2003). One of its advantages over other fields of inquiry is its fluidity. On the other hand, one of its disadvantages is its attachment to iconic theoretical representations of what the world ought to be. So, for example, graphical and mathematical models of optimal land use and location dominated the education of one generation of economic geographers to be replaced by theories of political economy that foretold the crisis of capitalism and, in particular, the ruinous consequences of global financial markets. From nineteenth century theories of land use and location came a preoccupation with the production of commodities and the distribution of those activities across the landscape – a preoccupation that remains evident today (given that the boundaries containing production have increased in scale from the region to the nation and now the globe, Swyngedouw 2000).

One of the crucial assumptions made by economic geographers over the past two decades is that the nation-state is, and arguably remains, one of just a few basic building blocks in analysing the economic geography of capitalism. So, for example, Clark (1980) used the Marxian notion of the reserve army of the unemployed in conjunction with related arguments about the role of the state to develop a model of regional differentiation built upon the social relations of production. Labour, capital, and the state were joined together to provide a model of industrial capitalism closely related to Galbraith's (1967) new industrial state. In retrospect, however, these analytical categories were quite narrow; labour meant the industrial working class and its unions, capital meant corporations and their productive assets, and the state meant the nation-state as regulator as well as its local representatives. This type of logic has been written out in economic geography and the social sciences in different ways; see, for example, Clark (1989), Esping-Anderson (1990) and Storper and Walker (1989). It is obvious that much has changed over the past 25 years even if, in some quarters, there is regret for the passing of an era (Jameson 1997).

Over the past 25 years, the methods and techniques of economic geography developed so as to give 'voice' to the political economy of landscape evolution. Ways of writing case studies and ways of portraying the exercise and negotiation of power between stakeholders have also evolved. If associated with economic geography, it has proven to be a mode of inquiry that resonates with finance (Jensen 1993) and with business and management (as exemplified by Wrigley 2000). But there are increasing problems with this kind of research strategy. Its production-led analytical categories are often misleading about the geographical scale at which decisions are taken and exaggerate the status of labour, capital, and the state in relation to new forms of finance capitalism operating at the local and global levels. To the extent that finance is recognized as an important driver behind industrial and regional restructuring, it is too often located 'offstage' shrouded behind curtains of ignorance of its principal imperatives and modes of practice. It is arguable that that which happens offstage behind those curtains is in fact the real stage of modern capitalism and should be, in actuality, the focus of economic geography.

In this chapter, I try to show how and why the geography of finance is so important for the future of economic geography. My goal is to enlist the interest and support of new generations of academics and practitioners concerned to better understand the structure and performance of global financial markets. Remarkably, the field of finance as an academic discipline is beginning to unravel as the theoretical axioms that held it together are exposed as inadequate in the light of the heterogeneity of practice. In doing so, I chart the rise of finance over the past few decades, the increasing gap between the theory and practice of finance, and a set of important research themes or problems that should engage economic geography over the next 25–50 years. Inevitably, my view about the future is partial: nevertheless, I hope the issues identified are sufficient to engage the reader in pursuing their own research on the geography of finance.

A world transformed

It is commonplace to look back over the past 100 years for crucial turning points or moments that mark the long-term structural transformation of modern economies (Webber and Rigby 1997). Just as Berle and Means' (1933) model of early twentieth century capitalism resonates today in studies of economic geography and corporate governance, Clark and Hebb (2004), Galbraith (1967) and Shonfeld (1965) set-out the logic of the Keynesian state and its relationship with managerial capitalism that remains important for economic geography. Nevertheless, it would appear that structural change has gathered pace over the past decades with a crucial turning point around the early 1970s marking the on-set of more than a decade of macro-economic instability. By the 1980s, corporate restructuring had become the single most important story about the transformation of the economic landscape, putting in play the role and status of whole sectors of industry, where they were located, and their relationship with what was now an independent financial sector (Clark et al. 2006). The 1990s overturned the world of production to become world of market speculation where, at one point, just one high-tech firm was worth more than half a dozen of the world's largest industrial conglomerates.

My argument in favour of the geography of global finance is about the importance of financial flows. It is also about the role of financial institutions that now occupy the centre stage of modern capitalism rather than being whispers heard offstage and outside of the intellectual purview of theorists. At each and every moment in the evolution of twentieth century capitalism, financial markets and institutions have played crucial roles in either prompting or facilitating economic change. This was certainly true during the 1920s and 1930s even if, in the years immediately after the Second World War, the political economy of corporate capitalism was deliberately conceived to constrain the influence of financial markets. Not only was the 1980s corporate restructuring fuelled by the burgeoning growth of financial assets, the high-tech bubble economy of the 1990s was made possible by massive inflows of financial assets to business angels through to the IPO market and beyond (Babcock-Lumish 2004). The intersection between the knowledge economy and financial innovation has become an important reference point for economic development policy in the European Union, Asia and Latin America.

The history of the nineteenth century can be written as a history of financial instability, economic insecurity, and the emergence of the nation-state (Bayly 2003). These elements of history can be woven together to create one meta-narrative such that the nascent nation-state gathered power and legitimacy over the first half of the twentieth century in the shadows cast by financial turmoil and the resultant economic insecurity. Even now, financial markets are greatly distrusted by many in continental Europe who associate 'irrational exuberance' with fascism and war (Clark 2003). The dominance of the nation-state in many western economies in the decades following the end of the Second World War can be traced, in part, to distrust of markets and the consequent absorption of financial and productive assets within the Keynesian state. Yet, at the time when the nation-state claimed a pivotal role in economic management, its powers (if not its legitimacy) were being surpassed by increasing volumes of international trade and the rise of a new kind of financial sector shorn of past alliances with the nation-state and with corporate capitalism. For those untutored in the debates of the 1960s, 1970s and the 1980s over the role of the state, the world of financial markets seems as natural as globalization.

There is a deep and intimate relationship between financial markets and globalization. Most obviously, the former makes the latter possible in the sense that trade through currency transaction is a necessary function for the exchange of commodities. But we also know that the daily volume of foreign exchange dealing is many, many times greater than that required by commodity trade; trillions of dollars are traded against a handful of currencies seeking momentary advantage in a system of relative pricing that shows little rhyme or reason if judged against whole countries' current and future prospects. A massive industry has developed around the arbitrage of currency risk and return where each and every trade begets a myriad of trades aimed at dispersing risk. Convention would have it that this kind of trade and the dispersal of risk contributes to financial instability. But it is equally possible that the pricing of tiny gaps in the market and the virtual and simultaneous spreading of risks across other traders and around the world actually contributes to financial stability. Commonplace stories of global financial instability do not take seriously the institutional management of risk around the world on a 24-hour basis (see Clark and Thrift 2004 and compare with Schoenberger 2000).

At the other end of the geographical scale, finance is increasingly the engine driving urban structure and differentiation. For many years, urban development was about the leverage of the public sector for private interest believing that the former was the most important source of investment in urban infrastructure – something that could not be priced or at least traded in the same way that currencies and commodities can be traded. However, it is increasingly apparent that governments are unable to provide the necessary volume and value of capital for new investment in urban infrastructure let alone regeneration of investments that, in many cases, go back at least a century if not two centuries past. In part, this is because of political constraints on the tax-raising capacity of national and local governments. In part, this is also because governments are increasingly caught between profound trade-offs such as expenditure on the education of the young and expenditure on the health care of the elderly (Clark 2000). Political constraints on tax-raising capacity have forced governments to rethink the virtues of financial markets, while financial institutions have become better able at pricing and distributing the risk associated with spatially fixed infrastructure investments. Whatever the consequences for equity, financial markets are quietly re-making modern cities (Babcock-Lumish and Clark 2005).

The nation-state as the banker of first call (smoothing the path of transactions, local and global), a presumption that informs much of the debate about the role of the state over the period 1950 to about 1980, is now a chimera. The nation-state and its representatives at the local level face an ever-tightening fiscal squeeze between competing generations' claims on limited tax revenue. The role of the state is as a regulator of financial markets and institutions, more focused upon ensuring capital adequacy and adequate mechanisms of risk-management than making up the difference where capital markets discriminate against certain groups in favour of other groups (local and global).

Principles and practice of finance

One of the most remarkable achievements of the academic discipline of finance has been its development and codification of commonly accepted theoretical principles. This is evident, for example, in the status accorded to the efficient markets hypothesis, the capital asset pricing model, and the Black-Scholes option-pricing theorem. These principles are the building blocks for introductory texts in finance, and serve as the reference points for new developments in the field that may or may not match the expectations framed by these principles.

The field is also highly quantitative drawing upon the most exotic mathematics through to the most recent developments in dynamic stochastic time-series modelling. In combination, papers published in journals such as *The Journal of Finance* and the *Journal of Financial Economics* are a clearly recognized genre.

But there is a paradox at the heart of the theory of finance: for all its logic and rigour the practice of finance is quite heterogeneous. This is all the more remarkable considering there is a vibrant market for investment analysts and investment managers that hold to theoretical principles, utilize the most advanced qualitative methods, and proclaim their commitment to best-practice in financial engineering. See, for example, the manifesto embodied in the Goldman Sachs approach to investment management (Litterman et al. 2003). At one level, for all the formal commitment to theoretical principles, it would appear that it is expected that the practice of finance is responsive to unanticipated events outside of the parameters set by those principles. At another level, however, it could be that theoretical principles are a charade in a world driven by successive waves of fashion and style amplified by the media wherein investment professionals legitimize their planned strategies if not their real actions by reference to these principles (Clark et al. 2004). The recent technology, media and telecommunications (TMT) boom and bust is an instance where observed behaviour went far beyond accepted principles of finance, justified in many instances by claims that the 'new' economy invalidated those principles (Shiller 2000, 2002).

Over the 1990s, the hegemony of the efficient markets hypothesis was such that any talk of so-called 'capital gaps' in cities, in emerging markets, or in any other market than traded securities markets was dismissed with a theoretical wave of the hand. The TMT bubble and bust brought into the open an empirical world not so easily rationalized against theoretical principles. It was recognized that the actual practice of finance is far more diverse and systematically different from first principles than theoreticians would have us believe possible. Since so much of economic geography is based on fine-grained knowledge of markets and institutions in particular times and places, the fact that the practice of finance is similarly empirically-driven is reason enough for economic geographers to study financial markets for their actual functions and performance. So, for example, there is a pressing need to study in some depth the reasons why finance is absent from some kinds of places at some times whereas finance may swamp other places at other times with enormous volumes of resources to the point of overwhelming global financial markets (as was arguably the case during the TMT bubble).

Another important feature of the practice of finance is its dependence upon hierarchies of tasks and functions as well as teams of individuals for the execution of investment strategies and the trading of financial products. One of the lessons learnt by large global financial houses during the 1990s was that individual traders are as much a liability as they are valuable stars in marketing efforts for new clients and new tranches of money to be managed. As we have shown (Clark and Thrift 2004), risk-management within financial houses and across the world on a 24-hour basis requires heavy investment in the monitoring and management of traders' risk exposures. Cadres of people must be employed at the interstices between markets so as to oversee risk exposures set against parameters managed by senior executives of the corporation. Not only is there a geography to risk-management, the hierarchy of tasks and functions within such complex organizations have quite distinctive ethnic and gender-specific labour markets: for example, compare the sources of clerical and semi-professional employees against the sources of highly educated investment professionals in the City of London or for that matter New York. In a nutshell, it is worth investigating why traders are men and why those that market trading functions to clients are women (see generally McDowell 1997).

In suggesting that there is a significant gap between the principles and practice of finance, and in suggesting that the practice of finance itself is subject to recognized social processes of differentiation and distinction, I echo arguments made by a number of social scientists (see, for example Knorr-Cetina and Preda 2004). At the same time, recognizing that at the core of finance there are well-established and observed customs and norms of research, the world of finance is open to the ideals and methodologies championed by economic geographers over last couple of decades.

Working from the bottom (the practice of finance) to the top (the theory of finance) provides us an opportunity to interrogate accepted principles. This has the virtue, of course, of building a conceptual understanding of the world of finance by induction rather than deduction. It also has the virtue of joining an increasing number of those in the finance industry who are seeking new ways of conceptualizing how financial markets work. Most importantly, the tools of economic geography can bring new insights about the structure and performance of financial markets given that the threads binding the field of finance together (such as the efficient markets hypothesis) are unravelling. At the same time, there remains scope for holistic models of the structure and performance of global financial markets, especially those that take seriously the interaction between markets and the geography of capital flows from the community through to the system of global circulation (Clark 2005). These models may require, however, the insights of close dialogue and the econometric techniques of large-scale data analysis (Clark and Wójcik 2005a, 2005b).

A research programme

By preference and by design, much of my research on global finance is empirical. This is because of my own interest in issues of public policy and the intersection I can see between social welfare and financial markets whether in the Anglo-American world, continental Europe, or beyond (Clark 2003). This is most obvious when considering recent work on nation-state pension systems and the imperatives driving global financial markets. But it can also be found in recent research on urban economic development and the pricing of economic landscapes. In setting-out a research agenda for the future, I have been conscious of the need to identify themes that have not only an empirical dimension but also strong and pressing theoretical currents. In the interest of brevity, four themes or problems are identified that have, I believe, sufficient scope to last economic geography for about 25 years!

One of the striking consequences of the TMT bubble and bust was the realization amongst financial theorists that behaviour is more interesting and more important than it was ever given credence by those that advocated the efficient markets hypothesis. Individual and collective behaviour drive financial markets, providing its random character, its moments of herd instinct and, at times, its momentum. This is a commonplace observation (although one that now has credibility when compared to 10 years ago). In terms of research, much of recent work has been on the intersection between decision-making and cognitive ability – the extent to which we are able to process information and respond in ways that are consistent with our self-interest. Equally important, but understudied, is the intersection between behaviour and context or what behavioural theorists sometimes refer to as the 'environment'. So important is this issue and yet so lacking in depth and theoretical bite that even in a summary description of the issue it is difficult to give specification to what is the appropriate scale of an 'environment' to be studied and how that 'environment' affects day-to-day and long-term decision-making.

By itself, a research agenda focused upon behaviour and environment would be one that resonates with all human geography whatever its particular focus. However, we can take the issue further into the world of political economy. Whereas much of the finance literature focuses upon the intersection between decision-making and cognitive ability, governments of all political persuasions are increasingly shifting responsibility for social welfare to individuals believing that their exercise of choice is both appropriate to our deference to individual autonomy and appropriate in terms of allowing people to seek their level of long-term welfare. Much has been written about this policy agenda, including those that see it as yet another instance of neo-liberalism. But it is also an issue of culture and society in the sense that how choice is framed is necessarily an issue of how people understand themselves in relation to others and their expectations of what is proper behaviour and what is not proper behaviour. In this respect, the interaction between culture and financial markets is one of the most important research questions of the coming 25 years.

Thus far, I have suggested that the research agenda for the geography of finance must take seriously the environment(s) and culture(s) in which people find themselves and to which they contribute in the sense of building a common world which we share. It could be argued that these issues are necessarily 'local' in the sense that the geographical scale is the lived world of individuals, their families, and their immediate communities. But this would be misleading or at least highly idiosyncratic in that people's everyday lives are being integrated from the bottom through to the top of the geographical scale – what we consume, where we consume, and how we consume (for example) are all part of a global marketplace for sustenance and social differentiation. In fact, the world of consumption, to pick just one crosscutting theme important in economic geography, is undergoing profound economic, social, and cultural globalization.

Culture is neither local nor is it autonomous; it is, like the management of financial markets across the world, a political project (Clark 2001).

So much of economic geography presumes that the urban landscape is properly the responsibility of the state. This is for a variety of reasons, including a concern for equity (a commitment to access and use) and efficiency (recognizing the lumpiness of urban infrastructure, for example). Just as importantly, there are those that believe that the urban landscape is a matter of social and political responsibility in that it is, after all, the level at which people live and work. When joined together with theoretical claims about the fixity of the urban landscape it provides economic geography with a rationale for excluding consideration of financial markets in the provision of urban public goods. While little studied or recognized as such, specialized financial institutions have entered the business of pricing the landscape and building the landscape. This is an issue of risk management in the sense that recognizing the fixity of urban infrastructure is an issue of designing financial instruments that can simultaneously capture value while distributing risk. These are very sophisticated financial instruments, more interesting and far more complicated than simply buying and selling traded securities on public markets. As the nation-state retreats further from the provision of public goods in the face of competing claims on limited resources, we must look more closely at the theory and practice of building cities through financial markets rather than avoiding financial markets.

This is an issue, of course, that relies upon the development and sophistication of financial institutions. It is something that has its own geography in that some of the most sophisticated players in this market are concentrated in just a few global financial centres such as London, New York, and (curiously) Sydney, Australia. Implied by this map of expertise is a map of global financial flows. While it is commonplace to recognize that daily global financial flows are massive and ever growing, we tend to ignore the origins and destinations of finance. Over the next 25 years, the vast reservoirs of financial assets in Anglo-American economies and to a lesser extent continental European economies will be directed at emerging markets in ways quite unlike the experience of the twentieth century. All this, of course, depends on building a global architecture consistent with insuring the security of those capital flows from origin to destination. Implied by this new world of finance is a world built upon common platforms for trading and managing those flows and common expectations about the nature and quality of disclosure and risk-management. For all the debate about path dependence and convergence, global finance finds, inevitably, differentiation and distinctions between jurisdictions inefficient.

These four themes or research projects for the future would take us from the individual to the global separately and, most importantly, together. If Western governments get their way, individuals will be asked to place bets on whole countries and the security of global financial flows for their retirement incomes. How they do that, with respect to the environment and culture in which they make those decisions, and the consequences of those decisions for urban economic development near and far, are issues to go to the heart of geography as a discipline and economic geography as a field of interdisciplinary study.

Conclusions

For much of its history, economic geography has been preoccupied with where commodities are produced, where they are traded, and where they are consumed. This kind of logic can be found in textbooks, in past and in current research, and in a barely articulated world-view that joins us together as a field of academic research and practice. Of course, commodities matter. This kind of reference point provides us a means of understanding where people live and work and how they joined together in institutions of action and negotiation. But for too long this kind of recipe for research has been blind to the role and significance of global financial markets, the rules of investment, and the imperatives behind successive waves of industrial and regional restructuring (Christopherson 2002). Just as geography matters, finance matters!

One of the unfortunate consequences of a single-minded focus upon the sites of production has been an uncritical commitment to describing the landscape in terms of its particularities as if each and every region is its own ensemble of untraded interdependencies. This has led us down a path that has been joined by other disciplines such as political science and sociology similarly committed to understanding capitalism in terms of its variety rather than its commonalities (see Hall and Soskice 2001 and compare with Strange 1997). In my own research, I have made contributions to this theme albeit in the shadows cast by the world of finance and its arbitrage processes (see, for example Clark 1989). My objection to this kind of research is not an objection to the fine-grained case studies that are so informative about the geography of production. My objection is to claims that the best economic geography is that which begins with local units, with production, and with a commitment to path dependence. Finance is all about pricing and trading in and out of path dependence, whatever its jurisdiction, and whatever it special circumstances. We might object to this fact of life, but we should do so in terms of its social consequences (Clark and Wójcik 2007).

The world of finance seems very different to that which we inherited. It has a different language, dictionaries of terms and concepts that seem, at first light, to be quite foreign (Clark et al. 2006). But as I have shown in this chapter, the dictionaries of terms are subject to their own internal tensions. Furthermore, the tensions between the theory and practice of finance provides us a way of understanding the processes that promise to remake the global economic landscape over the twenty-first century. We should be part of understanding global finance, its markets, and its social and economic processes. At the same time, as is the case in all social science disciplines, we have an obligation to build holistic models of the whole world.

Acknowledgements

This chapter was written in the light of collaboration with Terry Babcock-Lumish, Rob Bauer, Tessa Hebb, Lisa Hagerman, James Salo, Kendra Strauss, Nigel Thrift, Adam Tickell, and Dariusz Wójcik. Linda McDowell and the editors made comments on a previous draft. Funding for research related to the paper has come from a variety of quarters including the European Science Foundation, the

Rockefeller Foundation, the Social Science and Humanities Research Council of Canada, the UK Economic and Social Research Council, as well as financial companies and related industry groups. None of the above should be held responsible for my opinions.

References

Babcock-Lumish, T. (2004) *Beyond the TMT Bubble: Patterns of Innovation Investment in the US and the UK*, working paper no. WPG04-14, School of Geography & the Environment, University of Oxford, Oxford.

Babcock-Lumish, T. and G. L. Clark (2005) 'Pricing the economic landscape: financial markets and the communities and institutions of risk management', in J. Wescoat (ed.) *Prospects for Cities in the 21st Century* (forthcoming).

Bayly, C. (2003) *History of the Modern World*. Oxford: Blackwell.

Berle, A. A. and G. C. Means (1933) *The Modern Corporation and Private Property*, 1st edn. New York: Harcourt, Brace and World, Inc.

Christopherson, S. (2002) 'Why do national labour market practices continue to diverge in the global economy?: the "missing link" of investment rules', *Economic Geography*, 78: 1–20.

Clark, G. L. (1980) 'Capitalism and regional disparities', *Annals Association of American Geographers*, 70: 226–37.

Clark, G. L. (1989) *Unions and Communities under Siege*. Cambridge: Cambridge University Press.

Clark, G. L. (2000) *Pension Fund Capitalism*. Oxford: Oxford University Press.

Clark, G. L. (2001) 'Vocabulary of the new Europe: code words for the millennium', *Environment and Planning D: Society and Space*, 19: 697–717.

Clark, G. L. (2003) *European Pensions & Global Finance*. Oxford: Oxford University Press.

Clark, G. L. (2005) 'Money flows like mercury: an economic geography of global finance', *Geografiska Annaler B*, 87(2): 99–112.

Clark, G. L. and T. Hebb (2004) 'Pension fund corporate engagement: the fifth stage of capitalism', *Relations Industrielles /Industrial Relations*, 59(1): 142–70.

Clark, G. L. and N. Thrift (2004) 'The return of bureaucracy: managing dispersed knowledge in global finance', in K. Knorr-Cetina and A. Preda (eds) *The Sociology of Financial Markets*, pp. 229–49. Oxford: Oxford University Press.

Clark, G. L. and D. Wójcik (2005a) 'Financial valuation of the German model: the negative relationship between ownership concentration and stock market returns, 1997–2001', *Economic Geography*, 81(1): 11–29.

Clark, G. L. and D. Wójcik (2005b) 'Path dependence and the alchemy of finance: the economic geography of the German model, 1997–2003', *Environment and Planning A*, 37(10): 1769–91.

Clark, G. L. and D. Wójcik (2007) *The Geography of Finance*. Oxford: Oxford University Press (forthcoming).

Clark, G. L., T. Hebb, and D. Wójcik (2006) 'The language of finance', in J. Godfrey and K. Chalmers (eds) *The Globalization of Accounting Standards*, pp. 21–42. Cheltenham: Edward Elgar.

Clark, G. L., N. J. Thrift, and A. Tickell (2004) 'Performing finance: the industry, the media, and its image', *Review of International Political Economy*, 11: 289–310.

Dicken, P. (2003) *Global Shift: Reshaping the Global Economic Map in the 21st Century.* London: Sage.

Esping-Anderson, G. (1990) *The Three Worlds of Welfare Capitalism.* Cambridge: Polity Press.

Galbraith, J. K. (1967) *The New Industrial State.* London: André Deutsch.

Hall, P. A. and D. Soskice (eds) (2001) *Varieties of Capitalism: The Institutional Foundations of Comparative Advantage.* Oxford: Oxford University Press.

Jameson, F. (1997) 'Culture and finance capital', *Critical Inquiry,* 24: 246–65.

Jensen, M. J. (1993) 'The modern industrial revolution, exit, and the failure of internal control systems', *Journal of Finance,* 48: 831–80.

Knorr-Cetina, K. and A. Preda (eds) (2004) *The Sociology of Financial Markets.* Oxford: Oxford University Press.

Litterman, B. and the Quantitative Resources Group (2003) *Modern Investment Management: An Equilibrium Approach.* New York: Wiley.

McDowell, L. (1997) *Capital Culture.* Oxford: Blackwell.

Schoenberger, E. (2000) 'Management in time and space', in G. L. Clark, M. Feldman and M. S. Gertler (eds) *The Oxford Handbook of Economic Geography,* pp. 317–32. Oxford: Oxford University Press.

Scott, A. J. (2000) 'Economic geography: the great half-century', in G. L. Clark, M. Feldman and M. S. Gertler (eds) *The Oxford Handbook of Economic Geography,* pp. 18–47. Oxford: Oxford University Press.

Scott, A. J. (2004) 'A perspective of economic geography', *Journal of Economic Geography,* 4: 479–99.

Shiller, R. J. (2000) *Irrational Exuberance.* Princeton NJ: Princeton University Press.

Shiller, R. J. (2002) 'Bubbles, human judgement, and expert opinion', *Financial Analysts Journal,* 58(3): 18–26.

Shonfeld, A. (1965) *Modern Capitalism.* Oxford: Oxford University Press.

Storper, M and R. Walker (1989) *The Capitalist Imperative: Territory, Technology, and Industrial Growth.* Oxford: Blackwell.

Strange S. (1997) 'The future of global capitalism; or will divergence persist forever?', in C. Crouch and W. Streeck (eds) *Political Economy of Modern Capitalism. Mapping Convergence and Diversity,* pp. 182–91. London: Sage.

Swyngedouw, E. (2000) 'Elite power, global forces, and the political economy of "global" development', in G. L. Clark, M. Feldman and M. S. Gertler (eds) *The Oxford Handbook of Economic Geography,* pp. 541–57. Oxford: Oxford University Press.

Webber, M. J. and D. L. Rigby (1997) *The Golden Age Illusion: Rethinking Postwar Capitalism.* New York: Guilford Press.

Wrigley, N. (2000) 'The globalization of retail capital: themes for economic geography', in G. L. Clark, M. Feldman and M. S. Gertler (eds) *The Oxford Handbook of Economic Geography,* pp. 292–313. Oxford: Oxford University Press.

7 Economic geography and political economy

Ann Markusen

Economic geography: a synthetic, normative and policy-relevant field

The great strength of economic geography is its ability to study place by synthesizing insights from social science and natural science fields. As a trained economist, I have always envied this conceptual breadth and was drawn, like many others, to economic geography because of it. Although some economic geographers have tried to construct abstract theories that are uniquely economic geographic, as in notions such as spatiality and spatial scale, I have always found this impulse puzzling. Narrow, if elegant, reasoning plagues the social sciences and limits the usefulness of many of its branches (above all, economics), just as the disciplinary divisions in the natural sciences thwarted ecological analysis for many decades. When confronted with annoying anomalies like imperfect competition or less than full employment, economics dismisses them into peripheral fields (industrial organization, macro-economics) to protect its maximizing mechanics of scarce resources and unlimited wants, its theory of the firm and celebration of markets. Geography, unencumbered by such orthodoxies, offers scholars and policymakers a remarkable arena for harnessing the best of the sciences in service of understanding and shepherding change.

Because it is not as subject to limiting normative underpinnings, economic geography offers its students greater leeway to question institutions and ideologies than many of the fields upon which it draws. Natural science, even some social sciences, are constrained by methodological norms. Economics is dreadfully limited by its explicit individualism and its emphasis on efficiency as a single-minded social welfare goal. Thirty years ago, economists were taught that equity and stability were also key normative goals, but in the intervening years, these have shrunk in significance – equity is now chiefly ceded to sociologists.

For these reasons, economic geography has attracted thinkers and practitioners who want to work more synthetically and without the conceptual, methodological and normative confinement of its contributing sciences. In this chapter, I examine the intersection between political economy and economic geography in the second half of the twentieth century, showing how the two together have created room for work that was powerful, complex and at times successfully oppositional to the worst of capitalist spatial practices.

I first briefly review the rise, fall and resurgence of Marxist political economy in the past century. I then look at the seminal work of a number of geographers, economists and sociologists strongly influenced by Marxist thought – David Harvey (1973, 1985), Stuart Holland (1976), Manuel Castells (1977), Doreen Massey and Richard Meegan (1978, 1982), Barry Bluestone and Bennett Harrison (1982), Gordon Clark (1989), Michael Storper and Richard Walker (1989), among others – and show what they brought from it to economic geography in the 1970s through the early 1980s. These include an emphasis on working class/race/gender analysis, an understanding of the corrosive and uneven impact of capitalist development on cities and regions, the case for meso-economic analysis, acknowledgement of the role of contestation and struggle, an appreciation for institutions, especially the role of the state, and a commitment to research and advocacy in the interests of the exploited. I then illustrate the elements of a political economy-informed economic geography by reviewing 20 years of work on the military industrial complex. In closing, I address the continued synergy between the two fields.

The revival of political economy in the civil rights and Vietnam war era

In the United States, and also in Europe, grand historical events in the post World War II era produced social and political movements that revitalized and transformed elements of Marxist political economy as an alternative to the cold war battle between communist and capitalist ideologies and blocs. Nineteenth century Marxism had embraced the labour theory of value of the classical economists but built a theory that depicted owners of capital as exploiting labour and creating systemic tendencies towards overproduction and crisis in the economic system as a whole. The strange marriage of communism with totalitarianism in the Soviet Union, with its terrible human cost and slower economic growth, discredited Marxist theories of exploitation and visions of socialism, even though they inspired widespread, successful unionization and enduring political parties in parts of Europe, Asia and Latin America. The Marxist theory of crisis was given a new face with a Keynesian twist, the under-consumptionist view and the extraordinary practical success of Keynesian macro-economic policies from the 1930s to the 1970s. The successful recovery of the US, European and Japanese economies following World War II further diminished the draw of Marxist political economy, which fell into disarray in the 1950s and early 1960s.

Beginning in the 1960s, three broad social and economic phenomena led to a resurgence of interest in Marxist thought among social scientists and philosophers. First, in the United States in particular, the civil rights movement underscored the persistent racism and exploitation of the poorest urban and rural citizens, discrediting a 1950s view that a rising economy lifted all boats. The rapid movement of better-off white majorities to suburban and sunbelt regions created growing inner city malfunctions, while some rural regions continued to fall behind (e.g. Appalachia). New debates mushroomed on the causes and solutions for poverty in an explicitly urban and rural framework. Second, the United States involvement in the

Vietnam War set off a powerful, youthful movement against government policy that propelled intellectuals and students to search for better explanations of world economic and political disorder. Third, the increased integration of the world economy, accelerating in the 1970s and inducing plant closings and negative job growth in many more developed regions of Europe and the United States, produced movements of blue collar workers and communities that challenged mainstream economic and social thinking. Each of these fed into a rediscovery of Marxist political economy and its refashioning as American radical political economy and European counterparts.

Political economy meets economic geography

Economic geography as it evolved in the Anglo-American academy up through the 1950s bore few marks of Marxist influence. In the immediate postwar period, the birth of regional science as an inter-disciplinary enterprise of economists and geographers moved economic geography in the opposite direction, towards an empiricist spatial mechanics that stripped analysis of its social and political behaviour dimensions. But beginning in the 1960s, younger geographers, economists, political scientists, sociologists and urban planners, self-taught in Marxism while trained in rigorous deductive and quantitative methods, began to write oppositional analyses from many quarters. In the 1970s and 1980s, these were often fresh and surprising works, creating quite a stir among younger scholars. Disciplinary barriers fell as confident, politically concerned scholars across these disciplines learned of each others' work and began to converse, debate and use each others' insights.

It is impossible to do more than highlight a few of the more unique and important of these contributions here. The first round was more urban than regional. Geographer David Harvey's powerful *Social Justice and the City* (1973) presented a scathing critique of postwar urban/suburban structure and paved the way for dissident work, including his own brilliant geographical application of Marxist crisis theory in 'the spatial fix'. Sociologist Manuel Castels' *The Urban Question* (1977) improved on production-centric spatial theory by emphasizing social consumption as a key determinant of urban form, mediated by social organizing and conflict. Economist Ann Markusen's 'Class and urban social expenditures' (1976), an analysis of American suburbanization, emphasized class enclave-building as a driver.

A second wave addressed community and regional job loss in Europe and the United States. Geographers Doreen Massey and Richard Meegan's (1978, 1982) precocious work on corporate restructuring and its impact on workers and communities, closely paralleled economists Barry Bluestone and Bennett Harrison's *Deindustrialization of America* (1982) focus on the same phenomenon. Both teams offered analyses of the causes of plant closings and downward pressure on wages and tested them on individual industries; both formulated pro-labour action strategies and worked closely with unions and communities to implement them. Geographer Gordon Clark's *Unions and Communities under*

Siege (1989) offered a powerful account of how corporate spatial strategies created tough dilemmas for democratic unions, demonstrated with case studies.

On a regional scale, Pierre Vilar (1978) wrote a pathbreaking holistic Marxist-influenced analysis of Catalonian regional economic geography. Markusen (1987) offered an historical political economy of American regional development, expanded the Marxist conceptual framework for studying regional economies, and developed an institutional interpretation of American politics and political structure as it bore on regionalism. Geographers Michael Storper and Richard Walker developed a labour and technology-focused interpretation of regional uneven development in *The Capitalist Imperative* (1989).

These formulators of the political economy of place, informed by their rigorous social and regional science training but driven by pressing intellectual agendas linked to place-based political movements, brought economic geography a number of new tools. These scholars operated on a meso-economic level (Granovetter 1985; Holland 1976), rejecting macro-economic aggregates as too crude and micro-economic analysis as too individualized, abstract and stripped of institutional and political context. Spatial economies were to be studied on the basis of their industries and occupations, not just as amalgams of individual consumers, workers and firms. Formal and informal associations and networks among people were central to the analysis. They reintroduced the concept of social class, struck from the universe of important actors by economists, who substituted the abstraction 'labour' in its place. Not only did they embrace the broader dimensions of the concept of class (ideologies, associations, culture) but they added complexity by insisting that cross-cutting concepts of gender, race and ethnicity be considered co-equal in geographical analysis (Markusen 1979).

Many of these accounts incorporated the Marxist emphasis on historical materialism and strove to trace the origins of contemporary economic geographical issues in decades, even centuries of evolving political economy. This work placed contestation and class (and race, gender, etc.) struggle centre-stage, analysing evolving urban and regional economies as the product not just of capitalist or market dynamics but of success and failure of political movements. In order to do so, political and cultural institutions were brought into focus.

A hallmark of this body of work is its acknowledged commitment to constituencies and to scholarly-informed advocacy. The work of Massey and Meagan, Bluestone and Harrison, for instance, assumed an equity norm, focused on working class concerns, and wrote policy and action conclusions to their work that counselled organizing for change and concrete policy solutions, many of which were surprisingly incremental and some of which were embraced and won. Research topics are selected precisely because they address the large, often new issues and problems of the times, and researchers often worked with unions, community groups and others in formulating a research design and conducting it. Results are presented in academic forums but also in more popular forms for constituencies and the general public.

These diverse contributions played a major role in shaping economic geography in the ensuing decades. In addition to their conceptual and theoretical contributions, including their introduction of work from the various social sciences, they made important methodological contributions that are part of the economic geographer's toolkit today. Planner and geographer Erica Schoenberger, for instance, was the first to employ corporate interviews and write about the technique (1991). Planner and political scientist Annalee Saxenian (1994) pioneered ethnographic techniques to study the political and economic geography of two high tech regions. Both these contributions are fruits of the grounded, meso-economic approach. This school also kept alive a healthy scepticism and habit of respectful debate with each other and other tracks within economic geography (e.g. Lovering 1991; Markusen 1999; Martin 1999).

The political economic revitalization of economic geography had its blind spots, many of which it inherited or shared with other strains in the field. It was not particularly attuned to culture – the Marxist focus on material conditions made it difficult to acknowledge the role of culture and offered no very good tools for studying it. It was also not very good on ideology. It rather broadly rejected intellectual history in favour of dialectical material, and was rather unselfconscious about its own operation as an ideology. And, it was not particularly evaluative. Scholars who advocated everything from worker ownership to minute changes in policy were often naïve about the prospects for success and rarely reflected later on the success or failure of their prescriptions.

The geography of the military-industrial complex

Many, many wonderful contributions have been inspired by and followed on these works and others equally pioneering. Two generations of younger scholars have extended these analyses deep into the terrain of labour, corporate strategy, community development, gender, race, ethnicity, and developing countries' experience, among others. Precisely because this body of work is so grounded and institutional, it is easiest to demonstrate its synthetic character, its power to analyse, its normative stance and concrete achievements and failures by looking at its contributions in a single area. I use the case of the military industrial complex because it is one I know well and to which I have contributed.

The questions driving this research area are the following: what is the geographic distribution of military industrial and personnel activity, what drives its changing spatial configuration, which communities and constituencies benefit from and are hurt by it, and what can be done to curtail its negative consequences? Economic geographers had not worked on this topic at all up through the Vietnam era and its aftermath, though some quite accomplished economists, especially during the Vietnam War, had analysed the macroeconomic implications critically. It was during the 50 per cent peacetime increase under Reagan in the 1980s that American and British social scientists and geographers began to probe the spatial patterning of this important and unique sector (Crump 1989). In a historical account relying on contemporary corporate interviewing, Markusen et al. (1991) argued that the sunbelt phenomenon in the United States had been mis-interpreted; it was not just sunshine and low labour costs that impelled the uneven regional development of the United States in the postwar period, but the huge and enduring impact of government spending on industry and military bases, heavily skewed towards the south and west, and to a New England revitalized by diversified defence activities. Hooks (1991) studied the distribution of World War II military industrial capacity in particular and demonstrated a marked shift in the geography of American manufacturing and population as a result.

This shift, which separated military from civilian production and created huge, permanent enclaves of relatively transient military personnel on bases in remote places, rendered host communities quite vulnerable to the political business cycle and to the vagaries of American military policy. The damage to the American economy went deeper than that, argued Markusen and Yudken in their *Dismantling the Cold War Economy* (1992). The structure of United States industry, the character of lead firms in the sector, the lopsided development of the labour force (especially among scientists and engineers), and technological priorities mirrored this geographically uneven development. Working in Britain, geographer John Lovering documented many of the same tendencies at both the regional and industrial scales (1988, 1990b).

This body of work was poised for heightened interest at the end of the Cold War. Military-dependent constituencies feared the negative impact of United States defence budget cuts of 40 per cent (70 per cent for procurement) within a few short years, while others saw an enormous opportunity to use freed up resources for other purposes. The question was whether and how the re-use of resources might happen in the same communities and regions that had hosted defence-related activities. Peace activists understood that the pursuit of a new, diplomacy-intensive and peaceful foreign policy required that the nation worry about defence conversion on a local basis; otherwise, pork-barrel politics might string out military spending and military-led foreign policy. At the local level, unlikely coalitions of trade unions, peace activists, local economic development advocates and smaller defence contractors emerged hungry for an understanding of their predicament and what might be done about it.

Because the economic geographers who had done this research were working in the political economy tradition, it was easy for them to shift into more intensive, localized work in tandem with these coalitions. They had the skills to analyse corporate structure and strategy, labour skills and organization, technologies, the role of the state, local economic development, and the regional economy. They were comfortable working with these constituencies and were willing to address a larger public, at both local and national levels. Lovering in Britain and Markusen and her colleagues in the United States wrote a body of journal articles over a decade, but they also wrote popular accounts and op eds (e.g. Lovering 1990a; for a reflection on Markusen's team's ten years of work on defence conversion, including its wins and losses, see Markusen 2006).

Did it make a difference? Of course, the movement for a peace dividend and for permanently dismantling the bulk of the cold war weapons systems is currently in remission. Huge increases in military spending associated with the

Iraq war have created an umbrella rationale for continuing old programmes as well as funding new ones. But the work of economic geographers, I would argue, did make major contributions to the institutional and programmatic realization of a considerable peace dividend in the United States during the 1990s. First, it helped many communities understand their crises, identify and secure transitional assistance, and work with firms and unions to shift plants, bases, people and technologies into other activities. Second, because defence conversion required federal government involvement (since it was often the owner and always the consumer of military-related capacity), intellectual work successfully made the case for institutional innovations at the federal level and quickened the pace and quality of conversion. Third, critiques of existing labour programmes helped speed reform of worker displacement and retraining for defence workers, many of whom were older, specialized, and clustered in regions hard hit by defence cutbacks. Fourth, critiques of military corporate strategies in this era – mega-defence mergers and aggressive efforts to export – helped to rein in approvals of mergers and the more egregious proposed sales of high tech equipment to developing countries, whose resources were much better spent on building a civilian economy (Markusen 2006).

On many other social, political and economic fronts, the scholarly and outreach activities of political economy-informed economic geographers have made significant contributions to altering the trajectory of capitalist development.

The future of the fruitful marriage

The cross-fertilization between political economy and economic geography has been a fruitful one. Many of the economic geographic contributions – conceptual, methodological, empirical – that came from the post-civil rights/Vietnam period of academic ferment have passed on into the mainstream of the field. I have highlighted the synthetic nature of these – how insights that emerged not only from political economy but also from sociology and political science have enriched economic geography and added to its toolkit. As the disciplines have slipped back into their separate grooves, cross-disciplinary exchanges have dwindled, although they are still particularly powerful in fields like urban and regional planning, where I teach. Of course, intellectual work in the high season featured here was enlivened by an explosion of activism and diversity of popular movements. There are counterparts today (environmental, human rights, feminist, union, and anti-globalization movements), but they are not as linked, visible and successful. In the absence of these, academic work tends to fall back into normalcy and research into more esoteric intellectual veins. Abstraction is back in fashion, while grounded theory and empirical testing of theory is embattled. Few ask whether our work is having a real impact on the world or is read by or of interest to intelligent lay readers. If we said it in plain language, would they find what we do important?

I have pointedly written a 'history of economic geographic thought' essay. I am concerned that our field is no longer training our students in how the field has evolved or in linking that understanding to events and movements in the larger society. I encourage students and younger faculty to read the classics in the field and to spend time, preferably in seminars or with others, grappling with the diverse bodies of work within economic geography. We should all read broadly and know about advances in the other sciences and humanities, especially in this field where synthesis is so remarkably possible and badly needed.

References

Bluestone, B. and B. Harrison (1982) *The Deindustrialization of America: Plant Closings, Community Abandonment and the Dismantling of Basic Industry*. New York: Basic Books.

Castells, M. (1977) *Castells The Urban Question. A Marxist Approach*. London: Sage.

Clark, G. L. (1989) *Unions and Communities under Siege*. Cambridge: Cambridge University Press.

Crump, J. (1989) 'The spatial distribution of military spending in the United States 1941–85', *Growth and Change*, 20: 50–62.

Granovetter, M. (1985) 'Economic action and social structure: the problem of embeddedness', *American Journal of Sociology*, 91: 481–510.

Harvey, D. (1973) *Social Justice and the City*. Baltimore: Johns Hopkins University Press.

Harvey, D. (1985) *The Urbanization of Capital*. Baltimore: Johns Hopkins University Press.

Holland, S. (1976) *Capital versus the Regions*. London: Macmillan.

Hooks, G. (1991) *Forging the Military-Industrial Complex: World War II's Battle of the Potomac*. Champaign, IL: University of Illinois Press.

Lovering, J. (1988) 'Islands of prosperity: the spatial impact of high-technology defence industry in Britain', in M. Breheny (ed.) *Defence Expenditure and Regional Development*, pp. 29–48. London: Mansell.

Lovering, J. (1990a) 'The labour party and the "peace dividend": how to waste an opportunity', *Capital and Class*, 41: 7–14.

Lovering, J. (1990b) 'Military expenditure and the restructuring of capitalism: the military industry in Britain', *Cambridge Journal of Economics*, 14: 453–67.

Lovering, J. (1991) 'Theorising post-fordism: why contingency matters', *International Journal of Urban and Regional Research*, 15: 298–301.

Markusen, A. (1976) 'Class and urban social expenditures: a local theory of the state', *Kapitalistate*, 4: 90–111.

Markusen, A. (1979) 'City spatial structure, women's household work and national urban policy', *Signs: A Journal of Women in Culture and Society*, 5(3): 23–44.

Markusen, A. (1987) *Regions: The Economics and Politics of Territory*. Totowa: Rowman and Littlefield.

Markusen, A. (1999) 'Fuzzy concepts, scanty evidence, policy distance: the case for rigor and policy relevance in critical regional studies', *Regional Studies*, 33(9): 869–84.

Markusen, A. (2006) 'The peace dividend: moving resources from defense to development', in J. Brandl (ed.) *Common Good: Ideas from the Humphrey Institute*, pp. 75–85. University of Minnesota: Humphrey Institute of Public Affairs.

Markusen, A. and J. Yudken (1992) *Dismantling the Cold War Economy*. New York: Basic Books.

Markusen, A., P. Hall, S. Campbell and S. Deitrick (1991) *The Rise of the Gunbelt: The Military Remapping of Industrial America*. New York: Oxford University Press.

Martin, R. (1999) 'The "new economic geography": challenge or irrelevance?', *Transactions of the Institute of British Geographers*, 24(4): 387–92.

Massey, D. and R. Meegan (1978) 'Industrial restructuring versus the cities', *Urban Studies*, 15(3): 273–88.

Massey, D. and R. Meegan (1982) *The Anatomy of Job Loss: The How, Why and Where of Employment Decline*. London: Methuen.

Saxenian, A. (1994) *Regional Advantage: Culture and Competition in Silicon Valley and Route 128*. Cambridge, MA: Harvard University Press.

Schoenberger, E. (1991) 'The corporate interview as a research method in economic geography', *Professional Geographer*, 44: 180–9.

Storper, M. and R. Walker (1989) *The Capitalist Imperative: Territory, Technology and Industrial Growth*. New York: Basil Blackwell.

Vilar, P. (1978) *Cataluna en la Espana Moderna*. Barcelona: Editorial Critica.

8 The education of an economic geographer

Richard Walker

Rather than engage the entire field of economic geography over the last quarter century, I would like to reflect on my own pathway through the discipline. I hope this won't be seen as an indulgence, but as a way of putting flesh and blood on an epoch. But how to track a career? We all construct and edit continuously the narrative of our lives, seeking some semblance of order and justification for our motley existence. My course has zigzagged through several areas that, while not exactly a random walk, nonetheless presents some difficulties in drawing a neat trajectory line. The work of a life may not be wholly coherent, but still manifests certain principles of being a geographer and social scientist. An evident difficulty is that I am not simply an economic geographer. Still, there has been a long-standing commitment to political economy that has shaped everything along the way.

My undergraduate degree in economics was actually an accident since I had started out my course work concentrating in sciences, math and engineering. I still adhere to a scientific ideal for rational inquiry and explanation of the world, despite everything learned in the meantime about the frailties and fallacies of the scientific enterprise and about the role of mind, morality and human nature of science. The accident of economics turned into a devotion under the influence of a few teachers, most notably Joan Robinson, who came to Stanford at the behest of the student government in 1969. Robinson made the study of economics seem vital, as well as critical of the existing order (though what was wrong with conventional theory I still could not quite make out). I even started graduate school in Economics at Stanford, before quitting in disgust at the absurdity of the neo-classicism being drilled into us. That wariness about mainstream economics warned me from early on that economics is never enough. To Economic Geographers, I say we have to be in constant dialogue with other fields and problems, whether environmental, political, or sociological. We are always grappling with complex social systems. While the study of economics is a necessity in a capitalist world, it is never sufficient.

From that abortive beginning as an economist, I went searching the college catalogs for Environmental Studies programs (there were effectively none at the time) and stumbled upon the newly minted Geography and Environmental Engineering Department at Johns Hopkins University. When I arrived at

Hopkins in 1971, I hoped to pursue some kind of resource economics program. That misbegotten notion faded under the influence of David Harvey and Reds Wolman, who opened my eyes to the broader horizons of geography. Although David is seen as a Marxist above all, he was deeply steeped in British Geography and managed to transmit that affection to me without any formal drills. Harvey also introduced me to Marx's *Capital*, which we struggled through together. My economics and economic geography are still inescapably Marxist, though always open to extension and hybridization. After all, I was a Green before I was a Red. This may be why I am not usually cited as a classic Marxist Geographer like Harvey or his later student, Neil Smith.

I came to geography as an environmentalist owing to the influences of my youth in the Bay Area, a hearth of American environmentalism in the 1950s and 1960s. At Hopkins, my first piece of serious research was on a woeful reclamation project in Nebraska (which helped in its defeat) and the misuse of benefit-cost analysis to justify dams. The first iteration of my dissertation was an inquiry into the National Land Use Control Act, then under consideration by Congress (which spoke to my keen sense of personal loss in the paving of Silicon Valley, where I grew up). When the Act died and my draft proved boring, Harvey suggested I expand the first chapter, a history of suburbanization, into the whole thing.

When I went out on the job market in 1975, I was hired to teach environmental courses, not economic ones. The Chair at Berkeley, David Hooson, told me it would be the kiss of death among his colleagues to talk about economies or cities, so my job talk was on wetlands on the Chesapeake Bay, another project from graduate school. After being hired at Berkeley, I taught such courses as Water Resources, Open Space, and Population and Natural Resources. In those years, I wrote about the Clean Air Act, water projects in California, a Dow Chemical petrochemical complex, the logic of industrial pollution, and land use controls – all of which had an important element of economic analysis to them. Unfortunately, I bolted from environmental studies before the field took off. A wrong turn, perhaps, but it would lead me to economic geography.

My dissertation, The Suburban Solution (1977), had a great deal of economic geography in it. There were three main elements of analysis: the land market, business cycles, and class struggle. The first gave the immediate impetus to developers to push and pull the urban fringe outward; the second provided the larger impulse for property booms and development excesses; and the third explained the buy-off of the working class through consumerism and housing in the suburban context. What was missing, however, was any sense of the role of industry in the outward flux of the American city. I spun off a couple of articles on the logic of American suburbanization (e.g. 1981), but, unfortunately, never turned it all into a book – thereby being forever scooped by Kenneth Jackson's *The Crabgrass Frontier* (Jackson 1987); take heed, newly minted PhDs! As a result, I was never categorized as an Urban Geographer. Such are the vagaries of the disciplinary life.

I jumped into the field of economic geography in the early 1980s, thanks to visiting stints at Berkeley by Doreen Massey and Bennett Harrison and a spectacular group of students in City Planning and Geography at Berkeley, which included AnnaLee Saxenian, Meric Gertler, Erica Schoenberger, Kristin Nelson and Amy Glasmeier. The mass plant closures of that era in Britain and the United States were the catalyst to rethinking industrial location theory ('New Industrial Geography') – just as the urban crisis of the 1960s had influenced Harvey and others to rethink cities.

I began writing with Michael Storper, one of many amazing graduate students I have collaborated with, and we did a series of articles that culminated in *The Capitalist Imperative* (1989). That book was meant to be an answer to the neo-classical, equilibrium location theory that had ruled the roost since Walter Isard. It took on board seminal contributions by Lloyd and Dicken (1977), Doreen Massey (1984), Barry Bluestone and Bennett Harrison (1982), and Allen Scott (1988). But it rested on a wider foundation drawn from reading in economic theory, industrial history, and labor studies.

The major arguments of *The Capitalist Imperative* were of two kinds. On the one hand, it emphasized the dynamics of economic growth rather than the static allocation models of location coming out of the (Alfred) Weber tradition. Growth is driven by capital investment, strong competition, and pervasive disequilibrium. The model of growth was a Marx-Schumpeter-Keynes hybrid. At the same time, the model of 'geographical industrialization' rested on a firm basis in production, including technology, labor process and the division of labor.

It bothered me that Neil Smith (1984) and David Harvey's (1982, 1990) ideas on economy and geography gained such currency while our theme of 'geographical industrialization' was not widely taken up. Smith and Harvey kept to the realm of high abstraction of capital theory without ever descending into the nuts and bolts of production, meaning that they played loose and fast with industrial history and spatial patterns. The geography of production is so much more dynamic, varied and interesting than concepts like 'uneven development', 'spatial fix', and 'flexible accumulation' imply. My views on this have not changed much, as can be seen from my chapter on production in Sheppard and Barnes (2000).

In the 1990s, Michael Storper went on to collaborate with Allen Scott at UCLA, pursuing a dense regional analysis. I was less enamored of the liberalism of the New Institutionalism and its epigones such as Charles Sabel, Michael Porter, and Robert Putnam. Class conflict, capital accumulation, and state power were left out of the equation. Instead, I wrote articles on the failings of flexible specialization theory, on value theory, and on the economic role of technical change (1985, 1988, 1989, 1995a). I further developed my ideas about the division of labor in *The New Social Economy* (1992), written with geography's leading philosopher (also part economic geographer) Andrew Sayer. This was an occasion to rethink such large economic topics as the definition of services versus production, comparative industrial systems, business organization, and class formation (things I had begun writing about in the 1980s). The result was, again, somewhat disappointing in that our reflections intrigued readers but did not become a part of the collective imagination of economic geographers (let alone sociologists and the rest).

Instead, the decade of the 1990s saw me return to urban geography, using the San Francisco Bay Area as point of entry. I had been teaching Urban Field Geography since 1980, after taking over the course from Jay Vance. Despite tense relations with the prickly Vance, I learned a great deal about urban history from his writings. Allan Pred's historical work was another huge influence, and he became a good friend, as well. I've never lost my belief that without historical depth, economic and urban geography are inevitably shallow enterprises. This has often sent me into the arms of the historical geographers (who are another world apart), and made me skeptical of many of the glib claims coming from economic geographers about Post-This and the New-That.

This phase of work started with a long essay on the Bay Area (1990). It was inspired in part by Mike Davis's *City of Quartz* (1992) and Ed Soja's sweeping studies of Los Angeles (1989). I also had long admired Harvey's essays on nineteenth century Paris done in the 1970s (now out in a stunning book, *Paris: Capital of Modernity*, 2003). Mike asked me to turn the Bay Area essay into a short book for Verso Press, but, instead, the project exploded into a full-fledged attempt to capture the urbanization process in all its dimensions over a century. The idea was to combine the following:

- How industry molds cities over time (1996a, 2004a)
- How class and race divisions create a residential city of realms (1995b, 1996)
- How politics and social struggles over space have shaped the city (1998, 2007)
- How property development creates the built environment (1981, 1998, 2006)

These angles on urbanization combined several influences. The first was the reintegration of industrial location and city form. These had been sundered between economic and urban geography until Allen Scott put them back together in the 1980s. The second was how property development shaped the city, which Harvey (1973) had brought back into urban geography (and Harvey Molotch [1976] into urban sociology). The third was how class and class struggle shaped cities, which had been revived by Harvey, Chester Hartman (1984) writing on San Francisco's urban renewal, and Davis' political portrait of Los Angeles.

Another element – the look of the urban landscape – has been a significant part of my writing and teaching on the Bay Area (1995b). I firmly believe that in the distinctive elements of house types, gardens, and street layouts, among other parts of the built environment, one can find keys to the secrets of a city and a place. I never much liked the conservative views of J. B. Jackson, Pierce Lewis and other purveyors of the Landscape School in a previous generation; but my contemporaries in Cultural Landscape studies, such as Paul Groth (1994), Deryck Holdsworth and Gray Brechin (1999), have taken the field in very different directions. These are not names that regularly come up in economic geography, yet they have much to say about labor markets, merchant networks, office functions, resource flows, and more. The tensions between the old and new, left and right, in landscape studies are apparent in Groth and Bressi's collection, in which I have an essay (1997b).

Urban and historical geographers know that economic geography is never enough. It is only the skin and bones of cities and regions and countries, never the flesh. And the latter, the social order, is what gives places their face and their personality, and gives capitalism its necessary human and geographic form. Anyone coming out of urban studies doesn't need to rediscover local institutions, local governance, local cultures, and so forth in the way economic geographers have had to do; urban studies are inherently more attuned to politics, power, race, class and community, and less likely to fall into the traps of economism.

That necessarily means that my interest in the Bay Area has also been an extended inquiry into the social and political peculiarities of the place. On the economic side, this led to an inquiry into the character of California social relations and economic development going back to the Gold Rush (2001). My long look at California's social order took seriously Annalee Saxenian's challenge to economism in her study of Silicon Valley (1994), but pushed it much farther back in time than she was able to do – and made for a more ambiguous tale of the intertwining of regional social relations and regional economic development.

That project also grew out of a long dialogue with the 'roads to capitalism' approach to regional growth pioneered by Barrington Moore and Charles Post. It revisited some of the themes I developed with Brian Page (1991, 1994). We ruffled some feathers by challenging William Cronon's magisterial view of the region in *Nature's Metropolis* (1992), which, we argued, is just a variant of the Adam Smith trade theory of development, previously exposited by Vance (1970), that skips too lightly over the agrarian and industrial development of the Midwest (Cronon was not pleased, but we have since become very friendly, and he is publishing my latest book).

On the more political side, I tried to track California's contemporary condition (1995c). Without question, my view was darkened by the political malaise of the state and its anti-immigrant movement in the mid-1990s. Things turned around after that, but after another major economic crisis we've returned to reaction and degradation under Arnold Schwarzenegger. I became involved in resistance to Proposition 187 and wrote on immigration to California, including a pamphlet co-authored by Jeff Lustig (it was disowned by Mario Savio, leader of our little political coalition, because of objections by a couple of African-American members, before he and his son wrote a remarkably similar essay on their own). That experience, along with the creation of the American Cultures requirement at UC Berkeley, led me to plunge further into race theory and race history for my Geography of California course, and to incorporate racial order more thoroughly into my conception of class and political economy (1996b). A glimpse of these moves can be found in an essay in Roger Lee and Jane Wills' *Geographies of Economy* (1997a). They are the kind of necessary enrichment of social economic thinking we need more of in economic geography.

After what seems like forever – thanks to long interludes as a department chair and father – the Bay Area work will finally come together into two books on the urbanization of San Francisco and Silicon Valley (almost 20 years will have passed, making me feel rather old). In some respects, these are only particular case studies of American city formation. In other respects, the Bay Area is distinctive, as in what I've called its 'ecotopian middle landscape' of upper middle class residence or in its long history of maintaining the urban core as a cosmopolitan, politicized space. In still other respects, I've found the area a maddening combination of the unique and the mundane, like the juxtaposition of Silicon Valley's technical innovation and its banality of urban form.

A piece of the Bay Area project on the rural landscapes of the metropolis broke off to become a book of its own, a history of California agribusiness, *The Conquest of Bread* (2004b). This is a work of economic geography as much as anything else. As one might expect, key themes are the logic of agrarian capitalism, the expanding division of labor, production networks, class oppression, and the peculiarity of California's social order. These bump into secondary theses on remaking the natural landscape, the evolution of consumption, and so forth. Here, again, I was deeply influenced by two former students, Julie Guthman (2004) and George Henderson (1998), who have written brilliantly about California agribusiness, and Michael Watts, with whom I have shared many students in agrarian development. I also admire Don Mitchell's (1996) excoriation of rural landscape studies, though I depart from his narrowly farm-worker centered view of California agriculture.

The long tap root of my interest in agriculture goes back to the 1970s and my political education growing out of the movements supporting the farm workers and occupational health regulation. I didn't have to read agrarian theory to understand the importance of nature in agriculture, because I'd already been inculcated with the idea of real impacts of pesticides and water, among other things. And I carried that idea over to industrial geography in my treatments of technology, industrial variation and the labor process leading up to *The Capitalist Imperative*.

Another spin-off from the Bay Area project is a new book, *The Country in the City* (2007), on the way the countryside has become part of the urban fabric, especially as open space and parks. I argue for the distinctiveness of Bay Area environmentalism as a mass political movement and for the radical element of opposition to capital (especially property development). This historical geography takes me back to my political origins and highlights my own contradictory position as an upper class environmentalist and class-renegade friend of workers, immigrants and the poor. I am very likely too soft on white environmentalists, but the point is to show how important this kind of sustained critique of capitalism and American urbanization is – because it is so rare, so hard to maintain, and so much a part of a larger, reinforcing culture of left-leaning politics nurtured in what is known hereabouts as 'the Left Coast of America'. This project thus echoes the ideas about regional social order I have put forward with regard to economic development and the exploitation of nature, or ecotopian urban landscapes, but with a quite different twist.

Despite all the work on California, I do not believe in the priority of local over the global. I have to keep abreast of developments in global political economy for my course, the Economic Geography of the Industrial World. In the late 1990s, I wrote on the state of American labor in the face of global competition and global failure of capital accumulation (1999). Echoing Bob Brenner (2002) on the excesses of the 1990s, I argue that the fate of labor is not just about location and worker competition, but also the performance of national and world capital. On the other hand, I have debated Brenner about his relatively feeble approach to technology and geography. I have tried to link the global and the local in my latest paper on the influence of the US bubble economy of the late 1990s on the Bay Area and its urban landscape (2006).

So, in the end, my approach to geography is hard to put in a box. I regularly teach the global economy, but love to write about the local. I emphasize the grinding of the capitalist gears, but think that economic geography cannot make sense of things without social relations and politics. I see class all around but never doubt the significance of race, gender and nation. I watch with disgust the American empire trampling the globe in a thoroughly predictable way, while believing in the heroic achievements of a few dedicated Greens, counterculturalists or anarchists in the belly of the monster. I have tried to maintain my status as an iconoclast even as I've matured from Young Turk to Old Fart in the discipline. I have kept to my course, while being deeply influenced by brilliant people around me.

I have even come to terms with being a Geographer, with a capital 'G'. For a long time, I felt I'd backed into the discipline and could care less about the disciplinary obsessions of my colleagues. But time has worn down my contrariness. I've accommodated to being a Geographer. I see the discipline as in many ways better than the alternatives, like economics and sociology, which wouldn't know an ecosystem if it hit them in the eye. On the other hand, I do not believe that Geography is uniquely situated to know the world. What Geography does is to put me in contact with a lot of open minds and imaginative people who look at things in original ways. With time, I have come to see my career as a very long education of an economic geographer – though an education of a quite different sort than that of Carl Sauer (1963), whose essay title I've commandeered and whose long shadow of antipathy to things economic, political or modern hung over Berkeley geography for decades. So while arrived at by serendipity and circuitousness, the label Geographer will do as well as anything else. Economic Geographer sounds good, too – and is particularly useful when dealing with calls from the press, since no one in the United States seems to know what geography is. But just plain Geographer fits well enough to wear.

References

Bluestone, B. and B. Harrison (1982) *The Deindustrialization of America.* New York: Basic Books.

Brechin, G. (1999) *Imperial San Francisco: Urban Power, Earthly Ruin*, Berkeley: University of California Press.

Brenner, R. (2002) *The Boom and the Bubble: The US in the World Economy*. London: Verso.
Davis, M. (1992) *City of Quartz: Excavating the Future in Los Angeles*. London and New York: Verso.
Groth, P. (1994) *Living Downtown: The History of Residential Hotels in the United States*. Berkeley: University of California Press.
Hartman, C. (1984) *The Transformation of San Francisco*. Totowa, NJ: Rowman and Allenheld.
Harvey, D. (1973) *Social Justice and the City*. London: Edward Arnold.
Harvey D. (1982) *The Limits to Capital*. Blackwell: Oxford.
Harvey, D. (1990) *The Condition of Postmodernity*. Oxford: Blackwell.
Harvey, D. (2003) *Paris: Capital of Modernity*. New York: Routledge.
Henderson, G. (1998) *California and the Fictions of Capital*. New York: Oxford University Press.
Jackson, K. (1987) *The Crabgrass Frontier: The Suburbanization of the United States*. New York: Oxford University Press.
Lloyd, P. and P. Dicken (1977) *Location in Space*. New York: Harper and Row.
Massey, D. (1984) *Spatial Divisions of Labor: Social Structures and the Geography of Production*. London: Macmillan.
Mitchell, D. (1996) *The Lie of the Land*. Minneapolis: University of Minnesota Press.
Molotch, H. (1976) 'The city as growth machine: toward a political economy of place', *American Journal of Sociology*, 82: 309–32.
Page, B. and R. Walker (1991) 'From settlement to Fordism: the agro-industrial revolution in the American Midwest', *Economic Geography*, 67(4): 281–315.
Page, B. and R. Walker (1994) 'Nature's metropolis: the ghost dance of Christaller and Von Thunen', *Antipode*, 26(2): 152–62.
Sauer, C. (1963) *Land and Life*. Berkeley: University of California Press.
Saxenian, A. (1994) *Regional Advantage: Culture and Competition in Silicon Valley and Route 128*. Cambridge, MA: Harvard University Press.
Sayer, A. and R. Walker (1992) *The New Social Economy: Reworking the Division of Labor*. Cambridge, MA: Basil Blackwell.
Scott, A. (1988) *New Industrial Spaces: Flexible Production Organization and Regional Development in North America and Western Europe*. London: Pion.
Smith, N. (1984) *Uneven Development: Nature, Capital and the Production of Space*. Oxford: Blackwell.
Soja, E. (1989) *Post-Modern Geographies*. London: Verso.
Storper, M. and R. Walker (1989) *The Capitalist Imperative: Territory, Technology and Industrial Growth*. Oxford and Cambridge (US): Basil Blackwell.
Vance, J. (1970) *The Merchant's World: The Geography of Wholesaling*. Englewood Cliffs, NJ: Prentice-Hall.
Walker, R. (1977) 'The suburban solution: capitalism and the construction of urban space in the United States', unpublished PhD dissertation, Department of Geography and Environmental Engineering, Johns Hopkins University, Baltimore.
Walker, R. (1981) 'A theory of suburbanization: capitalism and the construction of urban space in the United States', in M. Dear and A. Scott (eds) *Urbanization and Urban Planning in Capitalist Societies*, pp. 383–430. New York: Methuen.
Walker, R. (1985) 'Technological determination and determinism: industrial growth and location', in M. Castells (ed.) *High Technology, Space and Society*, pp. 226–64. Beverly Hills: Sage.

Walker, R. (1988) 'The dynamics of value, price and profit'. *Capital and Class*, 35: 147–81.
Walker, R. (1989) 'Machinery, labour and location', in S. Wood (ed.) *The Transformation of Work?*, pp. 59–90. London: Unwin Hyman.
Walker, R. (1995a) 'Regulation and flexible specialization as theories of capitalist development: challengers to Marx and Schumpeter?', in H. Liggett and D. Perry (eds) *Spatial Practices: Markets, Politics and Community Life*, pp. 167–208. Thousand Oaks: Sage.
Walker, R. (1995b) 'Landscape and city life: four ecologies of residence in the San Francisco Bay Area', *Ecumene*, 2(1): 33–64.
Walker, R. (1995c) 'California rages against the dying of the light', *New Left Review*, 209: 42–74.
Walker, R. (1996a) 'Another round of globalization in San Francisco', *Urban Geography*, 17(1): 60–94.
Walker, R. (1996b) 'California's collision of race and class', *Representations*, 55: 163–83.
Walker, R. (1997a) 'California rages: regional capitalism and the politics of renewal', in J. Wills and R. Lee (eds) *Geographies of Economies*, pp. 345–56. London: Edward Arnold.
Walker, R. (1997b) 'Unseen and disbelieved: a political economist among cultural geographers', in P. Groth and T. Bressi (eds) *Understanding Ordinary Landscapes*, pp. 162–73. New Haven: Yale University Press.
Walker, R. (1998) 'An appetite for the city', in J. Brook, C. Carlsson and N. Peters (eds) *Reclaiming San Francisco: History, Politics and Culture*, pp. 1–20. San Francisco: City Lights Books.
Walker, R. (1999) 'Putting capital in its place: globalization and the prospects for labor', *Geoforum*, 30(3): 263–84.
Walker, R. (2000) 'The geography of production', in E. Sheppard and T. Barnes (eds) *Companion to Economic Geography*, pp. 113–32. Cambridge, Mass.: Blackwell.
Walker, R. (2001) 'California's golden road to riches: natural resources and regional capitalism, 1848–1940', *Annals of the Association of American Geographers*, 91(1): 167–99.
Walker, R. (2004a) 'Industry builds out the city: industrial decentralization in the San Francisco Bay Area, 1850–1950', in R. Lewis (ed.) *Manufacturing Suburbs: Building Work and Home on the Metropolitan Fringe*, pp. 92–123. Philadelphia: Temple University Press.
Walker, R. (2004b) *The Conquest of Bread: 150 Years of Agribusiness in California*. New York: The New Press.
Walker, R. (2006) 'The boom and the bombshell: the New economic bubble and the San Francisco Bay Area', in G. Vertova (ed.) *The Changing Economic Geography of Globalization*, pp. 121–47. London: Routledge.
Walker, R. (2007) *The Country in the City: The Greening of the San Francisco Bay Area*. Seattle: University of Washington Press (in press).
Walker, R. and the Bay Area Study Group (1990) 'The playground of US capitalism?: the political economy of the San Francisco Bay Area in the 1980s', in M. Davis, S. Hiatt, M. Kennedy, S. Ruddick, and M. Sprinker (eds) *Fire in the Hearth*, pp. 3–82. London: Verso.

9 On services and economic geography

Peter W. Daniels

Introduction

Service industries are everywhere, yet nowhere, in economic geography. Adam Smith and other classical economists promulgated the notion that service workers lack economic and social productivity. This has permeated writing and thinking about service industries and the economy to this day with 'the result that there is, scattered about in the literature, a sizeable number of writings that contain sharp insights but that, taken without careful analysis and interpretation, may appear confusing and even contradictory' (Stanback, Foreword to Delaunay and Gadrey 1992). Its impact is reflected in the visibility of services in contemporary economic geography; a quick scan of the contents list of some of the major recent collections of writings within the sub-discipline amply demonstrates the point (Barnes et al. 2004; Clark 2001; Sheppard and Barnes 2003). Apart from a seminal paper (Walker 1985) included in Barnes et al. (2004), papers directly focused on services as a distinguishable category of economic activity are notable by their absence. Yet they are almost certainly implicated in, for example, the numerous contributions to the discourses on realms of production, resource and social worlds, or global economic integration that populate each of these volumes.

Such an observation will probably be greeted by colleagues as yet another example of the paranoia among the small band of those interested in the economic geography of services about the neglect of these activities in mainstream research and writing. This phenomenon (if that is what it is) is not of course confined to economic geography; it applies in equal measure to urban geography, regional science, economics, or management and business studies. Given that this situation has been going on for at least 35 years (and probably longer) it may not serve any useful purpose to ponder the reasons, yet it might be argued that economic geography is the poorer for not explicitly incorporating service producing and service consuming activities within the philosophical, theoretical, methodological and other dialogues that characterised the sub-discipline during this period.

Another part of the services' identity crisis is the veracity of the distinction between manufactured goods and services. The widely cited observation by

Marshall (1961: 56) that 'Services and other goods which pass out of existence in the same instant that they come into it are, of course, not part of the stock of wealth' is noted. But Greenfield (1966) observes that later in the same treatise Marshall (1961: 63) states that 'Man cannot create material things' and concludes that 'all productive activities consist of services applied to pre-existing physical materials'. This is not too dissimilar to arguments and analyses made later by Riddle and others as the debate continued (Bacon and Eltis 1976; Cohen and Zysman 1987; Crum 1977; Riddle 1986). The absence of a clear cut distinction between goods and services is a recurring theme that imbues debates about the place of producer services in economic thinking, their relationship with the appearance of, and demand for, human resources with new and different skills, and a call for moving towards a new view of services (Daniels and Bryson 2002; Marshall et al. 1988). The term 'productive consumption' was used by Marx (1973) to represent the notion of significant interdependence between manufacturing and services; a product such as a washing machine, for example, only becomes 'a real product' when it is used for the purpose for which it has been designed (see also Bryson et al. 2004: 160–2).

Notwithstanding this evident lack of confidence, it is of course the case that there has been a long tradition of selective analysis of certain services, notably retailing, air transport, shipping and ports, tourism, office development and this has continued to the present but with different emphases; for example financial services, management consulting, advertising services or business and professional services more broadly defined. Research has also become more inter-disciplinary but, as a broad generalisation, each type of service activity is examined in its own right (operation, location, land use impacts, internationalisation) rather than as part of a common set of activities (services) that interact with, and exert influence upon, other common sets of economic activities (manufacturing, resource extraction, commodities production).

Given that 'service activities comprise very heterogeneous products, functions and occupations . . . *and* . . . a general approach would not result in meaningful explications' (Ochel and Wegner 1987) such fragmentation may be understandable and even inevitable. Nevertheless, during the last decade or so there has been a perceptible shift towards a more holistic approach (Bryson et al. 2004). Whether the stage has been reached where it 'is no longer necessary to begin any discussion of services research with a complaint about their neglect' (Marshall and Wood 1995) remains a moot point. At about the same time as the comment by Marshall and Wood it was also noted that 'in the universe of not-understood phenomena, service activities form one continent, only slightly explored from different angles' (Illeris 1996). This perceptive comment reflects Illeris's view that the youthfulness of service industry studies has precluded (*economic*) geographers from formulating a general theory of services; it has been necessary for them to work with, or draw upon, researchers in other disciplines to initially achieve a general understanding of services and their contribution to the transformation of economy and society in the late twentieth century.

The role of non-geographers

The validity of the observation by Illeris is borne out when trying to identify the sources of inspiration for research on services industries and the economy. Although one economist was moved to observe that the 'complete neglect of services in economic theory is almost incredible given the role of services in contemporary economies' (Hill 1977: 336), it can be suggested that much of the curiosity and pathfinding in services research has emanated outside economic geography. It can be traced to a number of seminal studies published by economists based at the National Bureau for Economic Research (NBER) during the 1960s (Baumol 1967; Fuchs 1964, 1965, 1968, 1969, 1980; Greenfield 1966) and to the work of Baumol (1967) who made a distinction between services where technology could be substituted for labour (thus improving productivity) and services where such substitution was not possible. The collective importance of these studies was their focus on services as a set of activities rather than as a single (uniform) sector separate from the manufacturing sector. There is no question that much of the thinking about the economics of service producing activities is derived from the book that pulled together all his earlier work (Fuchs 1968, for a summary of the reasons why it was such a pivotal analysis see Delauney and Gardrey 1992: 99–100). The relatively recent 'discovery' by economic geographers of the consumer as a factor of production, for example, was already appreciated by Fuchs who notes that 'productivity in many service industries is dependent in part on the knowledge, experience, and motivation of the consumer' (1965: 25, after Delauney and Gadrey 1992: 100).

Much of the interest in service industries and their productivity at this time was stimulated by their visible role in the economy and the changing employment structure of United States metropolitan areas (whether manifest as rapidly expanding office space, warehouses, department stores, or transportation services). Greenfield (1966) who was also working at the NBER used United States input–output data to extend understanding of the role of service industries in economic development; most importantly, he demonstrated inter-regional flows of service transactions and, in particular, the role performed by producer services. Several years elapsed before the significance of Greenfield's monograph was appreciated by economic geographers (Daniels 2001). Such is the significance of this work for one of the enduring research themes for economic geographers working on service services that it is worth elaborating on the contribution. Greenfield acknowledged that colleagues at the NBER such as Fuchs were already stressing the significance of the 'service economy' but that this equated with 'consumer services' rather than 'those services which business firms, non-profit institutions, and governments provide and usually sell to the producer rather than to the producer' (Greenfield 1966: 1). These are the 'producer services' that owe their existence to the 'direct purchase of services by one business firm from another' (Greenfield 1966: 2); they have since provided the focus for a small group of researchers, both within and outside the sub-discipline, committed to understanding their role in regional economic growth, the dynamics of world cities, or economic globalisation (Beyers 1989; Sassen 1994; Stanback 1979; Stanback and Grove 2002; Stanback et al. 1981; Stern 2001).

Greenfield himself devotes a chapter to an analysis of regional and industrial employment patterns in producer services in the belief that 'a deeper understanding of the economy and the conditions affecting economic growth can be gained through disaggregating the national totals (of employment) with the aim of discovering significant regional patterns' (Greenfield 1966: 93). Using a division of the United States into nine sub-regions, the regional concentration of producer services and the degree of inter-regional change for the period 1950–1960 is explored. The now familiar uneven distribution of producer services is clearly demonstrated; some 70 per cent of the (estimated) producer service jobs are located in just four regions. Input–output analysis reveals a significant relationship between the growth of producer services, the performance of the industries that they serve, and the degree of urban concentration. Explanations for variations in the regional distribution of producer services include localisation economies pulling producer services into metropolitan regions in a self-reinforcing process of cumulative causation; an analysis that has stood the test of time. The important point is that these and other topics, while they are only tentatively explored by Greenfield, have subsequently been very central to the producer services' research agenda for at least the last 20 years.

Not long after the group of economists at the NBER began to address services as integral to understanding economic development, sociologists were also beginning to make connections between changes in society and the rise of the service sector (Bell 1973). A shift from an industrial to a tertiary or service economy in the United States was already demonstrably under way in the early twentieth century as the share of workers in agriculture and in factories was steadily declining, to be replaced by a production system that is 'based on the co-ordination of people and machines' (Delauney and Gadrey 1992: 88) using scientific information and knowledge. This, in turn, promotes changes in occupation structure, such as the rise of the professional white-collar worker and the concomitant reliance on social networks for production rather than interactions between people and machines. From the perspective of this chapter, the significance of the writings by sociologists such as Bell and others with a similar disciplinary background (Browning and Singelmann 1975; Gershuny 1978, 1983; Sassen 1992) is that they provoked more careful thinking about the ways in which services as a fast-emerging category of economic activity were shaping production and, later, consumption. Some economists also explored the significance for economies of the evident decline in the share of manufacturing, often using the term 'de-industrialization' (Blackaby 1978; Bluestone and Harrison 1982). Others, such as Bressand and Nicolaidis (1989), Dunning (1989) or Miles (1993) extended the coverage to service and the global economy, the role of service multinational enterprises, or challenged the idea that service economies had replaced industrial economies; services were simply an expression of a new industrial economy. Political economists such as Petit (1986) and Hirschorn (1974) diversified the debate about the significance of the shift to services in relation to the drag they exerted on economic growth or the importance of earlier waves of labour and capital saving during the 1920s in the US that opened the door for new forms of work and societal development in the form of services.

Another theme which has greatly exercised those involved with services research is definition, classification and the related availability of suitably detailed/accurate statistics. The root of the anxiety can be traced, at least in part, to the uncertainty about the actual distinction between goods and services (see above). This is encapsulated in an early and very important paper by Hill (1977), an economist, in which the concept, definition and measurement of a service is elaborated in considerable detail. For Hill there is a major distinction between services affecting goods (such as the changes in the physical condition of goods by cleaning, repairs or decoration) and services affecting persons (changes in their mental or physical condition by activities such as education, entertainment, surgery, personal service, or communication) (see Hill 1977: 319–25). The outcomes from these two 'streams' of services may be permanent or temporary; a haircut brings about a temporary change in so far as it will need to be repeated while, other things being equal, surgery will generate a permanent change that is not reversible. It is also possible to distinguish, for both service 'streams' changes that are physical and changes that are mental, e.g. an entertainment experience or a sense of well being after a short holiday. Finally, some services such as education can be provided collectively as well as individually and can be distinguished from 'pure public services' such as fire, police and similar government services that 'Individuals are deemed to consume . . . all the time whether or not they want such services or are even aware of them' (Hill 1977: 338). The value of this analysis is its demonstration of the complexity of defining services and the knock-on effect on the potential for very complicated cross-classifications. Definitive solutions have yet to be produced; a group such as the Voorburg Group on Services Statistics (comprising national and international statistical agencies) are testament to this. Created in 1986 in response to a request from the United Nations Statistical Office (UNSO) for assistance in developing services statistics, the Group (statisticians, economists) involves exchanges amongst national statistical agencies and international organisations that lead to solutions of particular problems or the development of international guidelines in the field of service sector statistics. In recent years its agenda has expanded to include topics such as ways of estimating the real product of service activities, prices of service products, international trade in services, and employment, skills and occupations in the service sector.

Economic geographers and services

The contribution of scholars from other disciplines to the analysis and interpretation of service industries in economic development during the second half of the twentieth century has been considerable. The examples outlined above are by no means exhaustive and do not include research with roots in other disciplines such as business or management studies (Berry 1999) or urban and regional planning. Although the above hardly does justice to the range of work on services by other disciplines, it does provide a backcloth for examining the question: how have economic geographers contributed to research on service industries over the same period? The answer partly depends on whether their contribution is assessed as direct or indirect. Although such a dichotomy is vulnerable to the charge of over-simplification, there are undoubtedly numerous economic geographers who have used service activities such as retailing, tourism, transport, warehousing, research and development, or ecommerce as a means to an end: such as searching for explanations for changes in the organisation and location of production, the form and structure of cities, deindustrialisation, addressing problems in regional development, or understanding the changing relationship between consumption and production. Few, if any of them, would claim to be economic geographers with a curiosity about services. They have been, and continue to be, less interested in service industries *per se* as a category of activities and functions that, however heterogeneous, have together re-shaped ideas about how economies at scales from the micro to the macro evolve over time and the influence of services relative to the erstwhile drivers of growth and change: the manufacturing and primary sectors.

Some economic geographers remain implacably doubtful about services being anything other than subservient to manufacturing (or industrial) production. The best known is Walker (1985, see also Sayer and Walker 1992) who vigorously argue for an inclusive approach, whereby many services are only accessible or made possible by their incorporation within or justification through goods production. Some examples are transport, computer software, film, a consultancy report, or food outlets. The value of services is therefore dependent on material goods; this was possibly defensible a decade ago but is probably less the case today because certain information-intensive services such as computer software, film, publications of various kinds can be downloaded, stored and used without the need for storage on a material good such as computer disc. It is still necessary for a material good such as a computer to provide access to these services but, it could be argued, to a much lesser degree than before.

The 1980s and early 1990s surge

Although only a crude indicator, the number of books (authored or edited) by economic geographers solely concerned with service industries can be counted on one hand. Most have been published by economic geographers based in Europe, provide syntheses of some of the relevant international non-geographical and geographical literature, and reflect a shifting emphasis from a concern with theories of service sector development or the production and location of services during the 1980s and early 1990s (Bryson and Daniels 1998b; Bryson et al. 2004; Daniels 1985, 1991, 1993; Daniels et al. 2005; Illeris 1996; Marshall and Wood 1995). This does not stand comparison with the output from other disciplines and, although incorporating a 'geographical approach', it is probably fair to suggest that these texts have not shifted to any marked degree the agenda or the thinking by scholars or policymakers alike about the importance of incorporating or managing the part played by services in, for example, contemporary economic development.

This rather depressing conclusion must, however, be qualified in the light of some contributions by economic geographers that have challenged longstanding assumptions or highlighted significant developments that have reverberated beyond the discipline. A relatively early example is the research undertaken in the Puget Sound Region (Beyers 1986; Beyers and Alvine 1985) that demonstrated that services performed more than just a non-basic role (as widely assumed in the standard economic base model) in regional economic development; using an analysis of input–output data combined with a survey of individual service firms undertaken during the late 1970s, Beyers and Alvine were able to demonstrate that some 28 per cent of sales revenue was generated by transactions with clients in other parts of the United States and overseas. Furthermore, the share of non-regional sales revenue had increased from 18 per cent in 1958. The idea that some service industry output is tradable had been signalled by Greenfield (1966) but it had not been backed up with firm-level data. The Puget Sound Region research coincided with a number of similar studies in Europe that came to broadly similar conclusions. It also highlighted the fact that producer services are the key to understanding the basic function of services in urban or regional economies. Everyone recognises that many services such as personal, health or social services are largely non-basic but given that producer services have evolved into important sources of knowledge and expertise in the modern economy their share of total activity has expanded, making their tradability and availability a factor in the relative economic performance of places.

The 'discovery' of producer services provided a platform for the establishment in 1984 of the Producer Services Working Party (PSWP) with support from the Institute of British Geographers and the ESRC. It brought together a number of UK-based economic geographers as well as colleagues from other disciplines who set out to 'produce a state of the art review of research on producer services, examine secondary source evidence on their location and role, conduct a short research investigation into selected aspects of producer services, and to outline priorities for further research' (Marshall et al. 1988). Their documentation of many of the key aspects of producer services relied heavily on secondary sources and this only served to demonstrate how little was actually known about the dynamics of producer services, not least at the international level where their tradability was also an advantage. With secondary information widely acknowledged to be insufficiently disaggregated or lacking in coverage, the PSWP stimulated a whole new wave of investigations; it 'provided a means of thinking about service location . . . *but* . . . only described economic processes and the way they affect places in a very general way' (Marshall et al. 1988: 252). Nevertheless, its analysis of structural changes in the markets for producer services provoked a checklist that guided subsequent research, at least for much of the 1990s.

It included: identifying the strategic economic role of producer services in relation to production, consumption, and to other producer services; analysing in more detail the changing structure of demand for white- and blue-collar services; the basis for in-house versus externalisation decisions taken by different manufacturing, private, and public service firms; and the spatial implications of the reorganisation taking place in dominant firms (in finance, professional, and business services for example). The list goes on to include: the dynamics of the growth of small producer service firms (by far the most numerous in spite of the visibility of the larger, transnational firms); the types of producer services used as inputs at various stages of production (across all sectors), in the performance of various consumption functions or by customer organisations (in home markets or overseas); the processes of employment change, including gender issues and skilled versus unskilled human resources; the impact of information technology on the geographical disposition of service functions, on the demand for services, or on the nature of service organisation and delivery. Readers familiar with the research output of economic geographers undertaking research on services (in the sense used here) throughout the 1990s will hopefully recognise some relationship between the agenda set out by the PSWP and what actually happened (see for example Tickell 1999).

This is not to suggest that the group's work was the only source of inspiration (and indeed others may not see it this way at all) but it did help to focus the effort. Thus, in 1987 a group of French researchers (including a number of geographers) interested in pan-European comparative research on services and spatial development, but concerned about the scope for duplication of effort and a generally low level of research, convened a seminar in Paris to which they invited colleagues with cognate interests (from the UK, Italy, Denmark, Spain for example). The common sense of purpose engendered by the initial Paris meeting and later ones held in Lyon, culminated in the formation in 1988 of the European Research Network on Services and Space (RESER). At that time, very few researchers were involved in the conceptualisation (theoretical and empirical) of the role played by service activities, especially producer services, in regional or local growth. It has since expanded to incorporate 20 research groups or individuals (including economic geographers) active in services research and policy formulation located in 11 European countries (http://www.reser.net/). Via its annual conferences, publications (including annual reviews of services research in partner countries and selected papers from its meetings in the *The Service Industries Journal* and *Economie et Sociétés*), collaborative research activities (some funded by external bodies), or involvement of its members in major EU initiatives such as the Forum on Business-related Services (2004–5), the Network has kept alive the spirit of the PSWP while also adjusting its horizons to reflect changes in services research priorities at the start of the twenty-first century. Later initiatives, such as the North American Service Industries Research Network (NASIRN), unfortunately proved difficult to sustain.

Agenda diversification mid 1990s onwards

Another more general legacy of the PSWP (Marshall et al. 1988) did not really become evident until after the mid 1990s. Its message about the scale of the contribution of service activities to national GDP and competitiveness, their rapidly expanding share of employment and role as an alternative driver of change in cities undergoing serious industrial restructuring, or the burgeoning presence of services in processes of globalisation, international trade foreign direct investment, took a while to disseminate. But since the mid 1990s there has been a notable drilling down of the, still admittedly limited, scope of services research and writing by economic geographers. The enthusiasm for understanding the role performed by producer services in, for example, the global network of cities (Beaverstock et al. 1999) or in international business and trade (Bagchi Sen 1997; O'Connor and Daniels 2001) has been sustained and complemented with a plea for a more holistic view that incorporates consumer services (Williams 1997). At the same time there has been much more interest in exploring the characteristics and behaviour of particular sub-classes of services as a way of understanding their significance for spatial patterns of development or the socio-economic outcomes of, for example, the shift to service work. A few examples include computer services (Coe 2000), financial services (Bagchi-Sen 1995; Leyshon and Thrift 1997); cultural and media services (Beyers 2002a; Pratt 2000; Scott 1997), design services (Bryson and Daniels 2005; O'Connor 1996), or business and professional services (Bryson 1997; Daniels and Bryson 2005).

The other key dimension of services research by economic geographers during the last decade has been the emphasis on exploring processes mediated or shaped by inputs of the knowledge and expertise provided by these activities. Innovation by, and as a result of utilising, services has become a key contributor to corporate productivity and competitiveness in all sectors and in markets that are not only more global than ever before but also shaped by consumer needs and priorities (Howells 2002; Macpherson 1997). Knowledge-intensive business services continue to grow faster than most other activities in the economy; access to them by other firms and their role in the dynamics of the clusters within which many of them are embedded are crucial to the relative performance of local as well as national economies (Bryson and Daniels 1998a; Keeble 2002; Lindahl and Beyers 1999; Wood 2002a). Processes such as the changing nature and spatial redistribution of service work (Beyers and Lindahl 1996; Glasmeier and Howland 1995; Richardson 1999) also provide economic geographers with a rich vein for service-informed analyses of the changing balance of economic activity within and between national economies.

Conclusion

It has been suggested that even though economic geography must by definition embrace service industries within its research agenda it has largely neglected them as a category worthy of direct attention. Indeed, many of the lasting and significant theoretical and empirical contributions to our understanding of the role performed by services in the economy and development are attributable to scholars in disciplines other than economic geography. For activities whose behaviour is closely linked to factors such as accessibility, proximity, clustering and other geographical concepts this is perhaps surprising or perhaps a missed opportunity. Yet there are a number of milestones in the evolution of service industry studies involving groups of economic geographers or certain individuals that do represent a response to the general acceptance of the argument that services cannot be ignored. In this regard it is encouraging to note the recent attention devoted to charting future research directions, partly stimulated by the rise of the so-called 'new economy' (Beyers 2002b; Daniels 2004; Wood 2002b). One example will suffice.

Research on the relationship between developments in information and communications technology (ICT) and the supply, demand, quality and spatial distribution services is far from exhausted, not least as offshoring and outsourcing of both routine and higher-order service tasks presents economic challenges to some developed economies and opportunities for newly emerging economies. In addition, the widespread adoption by business and professional service (BPS) firms of ICT increases the potential for dispersal of the workforce and individualisation of work, including activities such as teleworking. It potentially undermines at least one of the rationales for the city. Their role as an intermediate source of knowledge and expertise is critical to sustaining local and regional economic performance; but their distribution is geographically uneven with significant concentration in regional cities with dispersed, low density patterns across city regions as a whole. With ICT increasingly mediating BPS production and distribution, as well as firm–client interaction, there is scope for established intra-urban and intra-regional location patterns of BPS to change over the next 10–20 years. Although the importance of face-to-face interaction and the need for BPS firms to complement underlying industrial specialisation still encourages agglomeration, some types of BPS are already becoming more dispersed, including home-based businesses. There is evidence that single-owner/SME production of BPS is important but the role of property and infrastructure in the ICT–BPS interface (especially for SMEs) and way in which economic and organisational factors interact to drive property and ICT demands in the BPS sector are not adequately understood and require more empirical research.

There have been very few occasions since the early 1980s when the annual meeting of the Association of American Geographers has not included at least one special session devoted to research on some aspect of services (the same cannot be said about the equivalent annual meeting of the RGS/IBG in the UK). The constituency has, however, been somewhat narrow and 'greying'; a relative absence in recent years of 'new blood' is a source of some concern as to whether the services dimension of economic geography is sustainable in the medium- to long-term.

References

Bacon, R. and W. Eltis (1976) *Britain's Economic Problems: too few Producers*. London: Macmillan.

Bagchi-Sen, S. (1995) 'FDI in US producer services: a temporal analysis of foreign direct investment in the finance, insurance and real estate sectors', *Regional Studies*, 29(2): 159–70.

Bagchi-Sen, S. (1997) 'The current state of knowledge in international business in producer services', *Environment and Planning A*, 29(7): 1153–74.

Barnes, T., J. Peck, E. Shepherd and A. Tickell (eds) (2004) *Reading Economic Geography*. Oxford: Blackwell.

Baumol, W. (1967) 'Macroeconomics of unbalanced growth: the anatomy of an urban crisis', *American Economic Review*, 57: 415–26.

Beaverstock, J. V., P. J. Taylor and R. G. Smith (1999) 'The long arm of the law: London's law firms in a globalising world economy', *Environment and Planning A*, 31(10): 1857–76.

Bell, D. (1973) *The Coming of Post-Industrial Society: A Venture in Social Forecasting*. New York: Basic Books.

Berry, L. (1999) *Discovering the Soul of Service: The Nine Drivers of Sustainable Business Success*. New York: The Free Press.

Beyers, W. B. (1986) *The Service Economy: Understanding Growth of Producer Services in the Central Puget Sound Region*. Seattle: Central Puget Sound Economic Development District.

Beyers, W. B. (1989) *The Producer Services and Economic Development in the United States: the Last Decade*, report to US Department of Commerce, Economic Development Administration, Washington, DC.

Beyers, W. B. (2002a) 'Culture, services and regional development', *The Service Industries Journal*, 22(1): 4–34.

Beyers, W. B. (2002b) 'Services and the new economy: elements of a research agenda', *Journal of Economic Geography*, 2(1): 1–29.

Beyers, W. B. and M. J. Alvine (1985) 'Export services in post-industrial society', *Papers of the Regional Science Association*, 57: 33–45.

Beyers, W. B. and D. P. Lindahl (1996) 'Lone eagles and high fliers in rural producer services', *Rural Development Perspectives*, 12: 2–10.

Blackaby, F. (ed.) (1978) *De-Industrialization*. London: Heinemann.

Bluestone, B. and B. T. Harrison (1982) *The De-Industrialization of America*. New York: The Free Press.

Bressand, J. and P. Nicolaidis (1989) *Strategic Trends in Services: An Inquiry into the Global Service Economy*. New York: Harper and Row.

Browning, H. C. and J. T. Singelmann (1975) *The Emergence of a Service Society*. Springfield: National Technical Information Service.

Bryson, J. R. (1997) 'Business service firms, service space and the management of change', *Entrepreneurship and Regional Development*, 9: 93–111.

Bryson, J. R. and P. W. Daniels (1998a) *Recipe Knowledge and the Four Myths of Knowledge-Intensive Producer Services Research*. Birmingham, USA: Service Sector Research Unit and Department of Geography, University of Bristol.

Bryson, J. R. and P. W. Daniels (1998b) *Service Industries in the Global Economy*, 2 vols. Cheltenham: Elgar.

Bryson, J. R. and P. W. Daniels (2005) 'Incorporating design services in contemporary economic geography', paper presented at the Annual Conference of the Association of American Geographers, Denver, April.

Bryson, J. R., P. W. Daniels and B. Warf (2004) *Service Worlds: People, Organisations, and Technologies*. London: Routledge.

Clark, G. (ed.) (2001) *Oxford Handbook of Economic Geography*. Oxford: Oxford University Press.

Coe, N. (2000) 'The externalisation of producer services debate: The UK computer services sector', *The Service Industries Journal*, 20(2): 64–81.

Cohen, S. and J. Zysman (1987) *Manufacturing Matters: The Myth of the Post-Industrial Economy*. New York: Basic Books.

Crum, R. E. and G. Gudgin (1977) *Non-Production Activities in UK Manufacturing*. Brussels: Commission of the European Communities.

Daniels, P. W. (1985) *Service Industries: A Geographical Appraisal*. London and New York: Methuen.

Daniels, P. W. (1991) *Services and Metropolitan Development*. London and New York: Routledge.

Daniels, P. W. (1993) *Service Industries in the World Economy*. Oxford, UK and Cambridge, US: Blackwell.

Daniels, P. W. (2001) 'Manpower and the growth of producer services', *The Service Industries Journal*, 21(4): 194–6.

Daniels, P. W. (2004) 'Reflections on the "old" economy, "new" economy, and services', *Growth and Change*, 35: 115–37.

Daniels, P. W. and J. R. Bryson (2002) 'Manufacturing services and services manufacturing: changing forms of production in advanced capitalist economies', *Urban Studies*, 39(5/6): 977–91.

Daniels, P. W. and J. R. Bryson (2005) 'Sustaining business and professional services in a second city region: the case of Birmingham, UK', *The Service Industries Journal*, 25(4): 505–24.

Daniels, P. W., K. C. Ho and T. A. Hutton (eds) (2005) *Service Industries and Asia-Pacific Cities: New Development Trajectories*. London: Routledge.

Delaunay, J. C. and J. Gadrey (1992) *Services in Economic Thought: Three Centuries of Debate*. Boston: Kluwer Academic Publishers.

Dunning, J. H. (1989) 'Multinational enterprises and the growth of services: some conceptual and theoretical issues', *The Service Industries Journal*, 9: 5–39.

Fuchs, V. R. (1964) *Productivity Trends in the Goods and Service Sector*, occasional paper no. 89, National Bureau for Economic Research, New York.

Fuchs, V. R. (1965) *The Growing Importance of the Service Industries*, occasional paper no. 96, National Bureau of Economic Research, New York.

Fuchs, V. R. (1968) *The Service Economy*. New York: Columbia University Press.

Fuchs, V. R. (1969) *Production and Productivity in the Service Industries*. New York: National Bureau of Economic Research.

Fuchs, V. R. (1980) *Economic Growth and the Rise of Service Employment*, NBER working paper no. 486, National Bureau of Economic Research, Cambridge, Massachusetts.

Gershuny, J. (1978) *After Industrial Society?: The Emerging Self-Service Economy*. Atlantic Highlands, NJ: Humanities Press.

Gershuny, J. (1983) *Social Innovation and the Division of Labour*. Oxford: Oxford University Press.

Glasmeier, A. and M. Howland (1995) *From Combines to Computers: Rural Services and Development in the Age of Information Technology*. Albany: State University of New York Press.

Greenfield, H. I. (1966) *Manpower and the Growth of Producer Services*. New York: Columbia University Press.

Hill, T. P. (1977) 'On goods and services', *Review of Income and Wealth*, 23: 315–38.

Hirschorn, L. (1974) *Towards a Political Economy of the Service Society*. Berkeley, CA: Institute of Urban and Regional Development.

Howells, J. R. L. (2002) 'Innovation, consumption and services: encapsulation and the combinational role of services', paper presented at the 12th International RESER Conference, Manchester, September.

Illeris, S. (1996) *The Service Economy: A Geographical Approach*. Chichester: Wiley.

Keeble, D. (2002) 'Why do business service firms cluster?: small consultancies, clustering and decentralization in London and southern England', *Transactions of the Institute of British Geographers*, 27(1): 67–90.

Leyshon, A. and N. J. Thrift (1997) *Money/Space: Geographies of Monetary Transformation*. London: Routledge.

Lindahl, D. P. and W. B. Beyers (1999) 'The creation of competitive advantage by producer service establishments', *Economic Geography*, 75(1): 1–20.

MacPherson, A. (1997) 'The role of producer service outsourcing in the innovation performance of New York State manufacturing firms', *Annals Association of American Geographers*, 87(1): 52–71.

Marshall, A. (1961) *Principles of Economics*, 9th (variorum) edn. London: Macmillan.

Marshall, A. and P. A. Wood (1995) *Services and Space: Key Aspects of Urban and Regional Development*. Harlow, Essex, England: Longman Scientific & Technical

Marshall, A., P. A. Wood, P. W. Daniels, A. McKinnon, J. Bachtler, P. Damesick, N. Thrift, A. Gillespie, A. Green and A. Leyshon (1988) *Services and Uneven Development*. Oxford: Oxford University Press.

Marx, K. (1973) *Grundrisse*. Harmondsworth: Penguin.

Miles, I. (1993) 'Services in the new industrial economy', *Futures*, 25(6): 653–72.

Ochel, W. and M. Wegner (1987) *Service Economies in Europe: Opportunities for Growth*. London: England.

O'Connor, K. (1996) 'Industrial design as a producer service: a framework for analysis in regional science', *Papers in Regional Science*, 75(3): 237–52.

O'Connor, K. and P. W. Daniels (2001) 'The geography of international trade in services: Australia and the APEC region', *Environment and Planning A*, 33: 281–96.

Petit, P. (1986) *Slow Growth and the Service Economy*. London: Pinter.

Pratt, A. (2000) 'New media, the new economy and new spaces', *Geoforum*, 31(4): 425–36.

Richardson, R. (1999) 'Teleservices, call centres and urban and regional development', *The Service Industries Journal*, 19(1): 96–116.

Riddle, D. I. (1986) *Service-Led Growth: The Role of the Service Sector in World Development*. New York: Praeger.

Sassen, S. (1992) *The Global City*. Princeton, NJ: Princeton University Press.

Sassen, S. (1994) *Cities in a World Economy*. Thousand Oaks: Pine Forge Press.

Sayer, A. and R. Walker (1992) *The New Social Economy: Reworking the Division of Labour*. Oxford: Blackwell.

Scott, A. J. (1997) 'The cultural economy of cities', *International Journal of Urban and Regional Research*, 2: 323–39.

Sheppard, E. and T. Barnes (eds) (2003) *A Companion to Economic Geography.* Oxford: Blackwell.

Stanback, T. M. (1979) *Understanding the Service Economy: Employment, Productivity, Location.* Baltimore: Johns Hopkins University Press.

Stanback, T. M. and G. Grove (2002) *The Transforming Metropolitan Economy.* New Brunswick, NJ: Centre for Urban Policy Research.

Stanback, T. M., P. J. Bearse, T. J. Noyelleand and R. Karasek (1981) *Services: The New Economy.* Totowa, NJ: Allanheld Osmun.

Stern, R. M. (ed.) (2001) *Services in the International Economy.* Ann Arbor: University of Michigan Press.

Tickell, A. (1999) 'The geographies of services: new wine in old bottles', *Progress in Human Geography*, 23(4): 633–9.

Walker, R. (1985) 'Is there a service economy?: the changing capitalist division of labour', *Science and Society*, 49(1): 42–83.

Williams, C. C. (1997) *Consumer Services and Economic Development.* London: Routledge.

Wood, P. A. (2002a) 'Knowledge-intensive services and urban innovativeness', *Urban Studies*, 39(5/6): 993–1002.

Wood, P. A. (2002b) 'Services in the "new economy": an elaboration', *Journal of Economic Geography*, 2: 109–14.

10 Towards an environmental economic geography

David P. Angel

When I first started teaching economic geography during the 1980s it was commonplace to begin an introductory class by asking the students to think about the geography of the clothes they were wearing. Students turned to each other and the tactile experience of pulling back the shirt collar connected the students personally to the course in a way the syllabus could not. The shoes 'Made in China' and the shirt 'Made in Sri Lanka' led easily into a discussion of economic globalization and of living and working conditions around the world. The students read the opening pages of *Global Shift* (Dicken 1986) and we were away. What is striking to me in retrospect is how little time we spent thinking about the resources from which the shoes and shirts were made, the chemicals used to turn a shirt sparkling white, the flow of waste water from the leather tanning factory, and the energy and pesticides expended in cotton fields. The material foundations and environmental consequences of economic activity, flows of energy, water, materials and waste, were if anything, less visible for me and the students than the social conditions and geography of production and consumption.

Nowadays undergraduate students are eager to discuss at least some environmental and material aspects of economic activity. Instead of asking where the iPod is made, we have an interesting discussion of whether or not the battery is recyclable, and what to make of Apple's adoption of a take-back policy on 'old' iPods. Is there an economic rationale behind the take-back policy and how does this rationale fit into the social regulation of business and into the market strategies of firms in the contemporary political economy? We talk a bit about the use of heavy metals in electronic equipment and the significance of the European Union RoHS directive.[1] We debate whether tough environmental regulation of economic activity stimulates innovation, or whether environmental regulation slows economic growth. Sometimes we make it to a discussion of energy intensity, greenhouse gas emissions and climate change and perhaps to thinking about the economic impacts of pollution. But we do not find much of this in the current textbooks of economic geography. Where is the 'environmental economic geography' equivalent of Global Shift? Alternatively, we might ask whether the above topics are of economic geography at all, or part of some other disciplinary or inter-disciplinary field of inquiry.

One of the striking elements of teaching economic geography in the 1980s was the easy alignment of topics taught with concepts and theories of the field. Students eager to engage with theory found a blossoming economic geography literature on industrial districts, commodity chains, agglomeration economies and the like that spoke directly to issues of concern, whether it be the loss of manufacturing jobs in OECD economies or the rapid rise of the Asian newly industrializing economies. More generally, these meso-level concepts nested well within broader theories of political and economic change, such as regulation theory. Stated differently, the focus and scope of theoretical work underway within economic geography was well aligned with the kinds of research questions that were in play and also connected well with some of the major public policy debates of the time.

In retrospect, this alignment of the topics and the theory of the field in the 1980s was constraining as well as it was enabling. As detailed in many of the other chapters of this volume, the past two decades have seen both acceptance of a broader definition of economic geography (symbolically marked in the United States by a shift from industrial geography to economic geography), and by a proliferation of alternative theories and epistemological standpoints. In some instances, such as that of feminist analysis, these new directions in economic geography have had a transformative impact on the theory and methods of the field. In other cases, such as the interest in finance capital, existing theories and method have proven adaptable to the issues of concern. Recent reviews of economic geography scholarship have stressed the polyvalent character of the field today and the added value generated by diversity of method and approach (Clark et al. 2000). As discussed later in this chapter, one consequence of this broadening of the concept of the 'economic' has been the engagement of economic geography with research that maintains an emphasis on human-environment relations, an emphasis that was largely absent from industrial geography during the 1980s.

But what has been the experience of researchers practicing environmental economic geography? As efforts are made to recover the material foundations and environmental consequences of economic activity, what happens to the theories and methods that we use and teach as part of economic geography? In this chapter, I reflect on the growing engagement of economic geography with issues of the environment. My central interest is with the theories that we have at hand to study economic geographies that are as much about flows of energy, material and waste as they are about flows of capital and goods produced, and about measures of success that relate to the energy intensity of economic activity as well as to livelihoods maintained, jobs created, and profits made. The chapter is bound by my own intellectual path dependency in that the discussion is limited to industrial activity.

I describe two principal approaches that have emerged within the field. The first approach is labeled the greening of industry. This approach in essence takes much of the existing conceptual apparatus of economic geography, from studies of commodity chains to work on innovation and technological change, and applies this work to issues of resource use and the environmental impact of economic activity. I argue that this approach aligns well with the actual practice of many industrial firms today. It is also an approach that is fundamentally firm-centered. The second approach, that is labeled a political ecology of industrial change, is far more nascent within the field, and largely absent from the economic practice of industrial firms. Here the theoretical and empirical frame of analysis shifts to a structural form. The fundamental question to be addressed is the ways in which the material foundations of economic activity – from resources used to waste generated, along with attendant environmental impacts – impact patterns and processes of industrial change. As the reference to political ecology suggests, however immutable biophysical processes and geochemical cycles might be, the engagement of economies with the environment is fundamentally a social process that requires interrogation through social (and economic geography) theory.

How did we get here?

For a long time the received discourse of geographers was that we were a discipline divided into physical geography and human geography. After a while, this discourse was clarified in terms of the identification of two axes of interest, one being human–environment relations, and the other being space and place (Turner 2002). In the case of economic geography, it is fair to say that concern with space and place has been the dominant strand of enquiry (Scott 2000). The concepts that economic geographers have brought to the fore, and the claims that we have made regarding the contribution of economic geography to the study of economic change, have been almost entirely to do with the spatiality of economic activity, whether this be the location analysis of the 1960s, or the analysis of agglomeration economies, industrial districts and globalization of the 1980s and beyond. This tendency to focus on space and place was reinforced by the primary focus of economic geography on industrial activity, and within industrial activity upon the manufacturing and service sectors that were dominant employers in OECD economies. Studies of domains and places of economic activity with more visible connections to the environment, such as subsistence agriculture in the developing world or oil and gas extraction in the Middle East, Russia and the North Sea, took place outside of an economic geography focused squarely on manufacturing and service industries. Even in resource-intensive manufacturing industries such as metal production, the primary interest of economic geographers has been the location of these production facilities rather than economy-environment relations.

This placement of economic geography squarely on the society-space axis of the discipline mirrored of course the externalization of issues of environment and resources within the activities of industrial firms themselves. While obviously varying tremendously across industrial sectors, resource use and environmental impacts were for many firms and industries of secondary concern to more 'central' business challenges, such as how to reduce labor costs and enhance labor productivity, how to accelerate the development of new products and to break into new markets, and how to manage increasingly complex global business organizations. Studies of industrial sectors such as food production, fishing, and primary resource industries, where issues of resources and environment were of central concern to business practice, sat uneasily on the fringes of industrial geography and rarely impacted theory development in the field.

Even when issues of environment and resources were thrust squarely in the forefront of economies, such as on the occasion of the oil price shocks of the 1970s, and the Bhopal chemical disaster of the 1980s, scholarship on these economy-environment concerns within geography tended to take place in specialized niches, such as study groups on energy and the environment, and on risk-hazards, rather than being a central and integrated part of the scholarship, and especially the theory development, of economic geography. Calls by Fitzsimmons (1989) and others for economic geography to address issues of environment and nature-society relations had little impact. In part this was a result of the institutional fragmentation that had been created within geography. As Castree (2004: 80) has noted '. . . because the environment is strung out between many different parts of human geography, it is difficult to generate a critical mass of researchers working on the same environmental issues, asking similar questions or deploying similar theoretical apparatuses.'

For all of this, during the 1990s and with increasing momentum over the past five years, issues of resources and the environment began to work themselves onto the stage of economic geography (see, for example, Angel 2000; Bridge 2002; Gibbs 2002). In part this was a direct result of a broadening of the concept of the economic and the engagement with different theoretical perspectives beyond that of the dominant political-economy of the 1980s. With increased interest in issues of development has come an engagement with an important literature on political ecology (Peet and Watts 1996). Understudied sectors, such as the processed food industries (Murdoch et al. 2000), have been a platform for important work on constructions of nature and on concepts of quality. Studies of other economic activities, from the tending of garden lawns (Robbins and Sharp 2003) to biotechnology (Marsden et al. 2003), have raised important questions about the environment and technology in everyday life. Economic geographers have also engaged with inter-disciplinary research on issues such as human dimensions of global environmental change (O'Brien and Leichenko 2003) and ecological modernization (Gibbs 2003). The concept of nature itself has undergone close scrutiny in ways that connect to the cultural turn in economic geography (Castree and Braun 2001). There has also been direct engagement with some important resource issues, such as water supply (Bakker 2004). Indeed, the literature of economic geography is now replete with calls to integrate the analysis of resources and the environment into the core of human geography (Castree 2004), economic geography, and urban geography (Braun 2005).

It is with firms and industries themselves, however, that the real driver of the current interest among economic geographers in economy-environment relations lies. To be sure, issues of environment and resources are now firmly and visibly part of the contested dynamic of industrial change for firms and industries, whether this is in the context of global climate change, health risks or water supply. But what is most distinctive about the current period is not so much the elevated interest in the environment as the attention paid by firms themselves to issues of resource use and environment impact in managing fundamental processes of investment, technological change and the organization of business enterprises. Innovation and learning, subcontracting relations, and supply chains are being examined from the perspective of environmental performance and resource use in a way that was far from common even a decade ago. Concepts such as design-for-the environment, ISO 14000 certification, life-cycle analysis, resource use score-cards, environmental footprints, are now common place tools of business practice among large firms. At the same time, the relation between environmental quality, resource use and economic change is now of central concern, not just to environmental regulatory agencies, but also to institutions responsible for economic development, ranging from Ministries of Industry to multi-lateral development agencies, such as the Asian Development Bank and the World Bank (see World Bank 1999). Most fundamentally, there is beginning to emerge some empirical evidence that efforts to meet environmental regulations and associated mandates are causing firms to rethink the overall organization of production systems, including the structure of global production networks (Angel and Rock 2005).

What underlies this more direct engagement with issues of the resources and the environment on the part of firms and industries? Several factors are in play. First, there has been important evolution in patterns of environmental regulation of firms and industries, and an increasing amount of self-regulation and adoption of voluntary compliance codes of conduct by firms seeking to avoid more direct regulation of their activities. Of particular importance is the growing significance of end-market regulation in which firms are regulated not just in terms of environmental impacts and resource use at the site-of-production but also in terms of the environmental profile of products sold. The most significant examples of such end-market regulations are the European Union RoHS and WEEE legislation that mandate the elimination of certain toxic substances from products sold in member countries along with minimum standards on product recycling. Because these end-market regulations address the material characteristics of products sold, they reach deeply into the operation of supply chains and production networks in ways that traditional site-of-production regulation does not. An electronics manufacturer needs to be concerned about whether the printed circuit board manufactured by a second tier supplier used lead solder in its production process, and what chemicals were used as solvents in production.

Second, firms and industries now find themselves subject to a higher level of scrutiny of their activities around the world, and scrutiny by a wider array of actors. In part this is simply a consequence of globalization and the increasing connectedness of places and people around the world. Activists now have the capacity to create networks of advocacy linking local communities in places such as Vietnam to shareholders in the United States. At the same time, institutional investors have become increasingly interested in the risks that weak environmental performance might present to shareholder value. In and of themselves, these developments might have elevated the pressure on firms to address issues of environmental performance, but they are unlikely to have had the more far reaching impact on basic processes of investment and technology change observed in some global firms today. What appears to have been decisive in leading firms to respond to these advocacy networks is the fundamental importance of 'reputational capital' to the long term competitiveness of firms in many sectors of the contemporary economy. This was perhaps most visibly evident in the vulnerability of name brand apparel manufacturers to pressures to improve working conditions in sweatshops. Where brand reputation is a crucial constructed asset, and where environmental health and safety concerns are issues that can potentially weaken the brand, firms are likely to be more responsive to advocacy networks.

Beyond these two factors, the more general concern voiced around issues such as climate change, as well as local and regional environmental challenges, has contributed to the pressure on firms to improve their environmental performance. This has been especially the case in rapidly industrializing economies in East Asia and elsewhere where the sheer scale and pace of industrialization and urbanization has placed extraordinary stress on environmental conditions and resources.

What to look for in an environmental economic geography

If economic geography students were to investigate the ways in which environmental and resource concerns are impacting the operations of industrial firms today, at least among many large multi-national firms, they would find a variety of significant activities underway, ranging from environmental supply chain management to investment and technology change designed to enhance energy and materials efficiency of economic activity and improve environmental performance. Many firms are now setting specific quantitative goals for enhancing energy and materials efficiency and reducing waste flows. What has been called the 'greening of industry', the efforts of individual firms and industries to improve environmental performance, provides one important focus for an environmental economic geography. Research in this area remains quite preliminary with even such basic questions as the scope and scale of greening underway, and the significance of these activities for overall environmental performance, very much in question. At the same time, many of the challenges that firms are dealing with, such as how to foster learning and innovation around energy and environmental performance in a large, complex production network, are at the heart of some of the contemporary debates in economic geography on learning networks, supply chains, and technological change. Indeed, it is quite striking the degree to which the analysis of the greening of industry draws easily and with real value upon theory and conceptual insights that have been generated by existing research in economic geography.

Three topics illustrate the kinds of research questions that economic geographers might engage with within the domain of 'greening of industry'. First, and as previously indicated, one of the key issues that has emerged for firms seeking to improve environmental performance is the management of supply chains and production networks within large business enterprises. Firms are now attempting to manage global production networks not just from the perspective of cost, quality and time-to-market, but also with respect to environmental performance. There is beginning to emerge some initial evidence to suggest that the introduction of environmental performance into the management calculus of firms is impacting the organization and geography of production networks, focusing production on a much narrower set of suppliers who can meet the environmental performance requirements of the parent firm. Concerns are being raised in developing economies that the environmental requirements being introduced into supply chains may foreclose some of the opportunities for industrial upgrading that have supported poverty-reducing economic growth (Rock and Angel 2005). Economic geographers are well placed to examine the likely consequences of these changes in global production networks for economic development at the regional and national scale. That is to say, the significance of the greening of industry goes beyond environmental impacts per se to studies of the ways in which efforts to improve environmental performance are impacting the organization and geography of production networks.

A similar cross-over between issues of environmental performance and existing research foci in economic geography is observed around issues of innovation, learning and technological change. In attempting to meet requirements to improve energy and resource efficiency, eliminate toxic substances from products and production processes, and reduce waste and emissions, firms practice learning, innovation and technological change. How these processes of learning and innovation take place, and in particular the geography and spatiality of these processes, has been a longstanding area of research focus for economic geography. In the case of innovation around environmental performance, much of the activity of firms is structured by either actual or anticipated regulatory requirements. This has led to a growing literature seeking to assess whether tough environmental regulation stimulates innovation and improved economic and environmental performance. But there are other dimensions of this linkage between environmental regulation and innovation. For example, partly in response to the difficulty of assuring compliance with multiple different country-based environmental standards, some large multi-national firms are adopting firm-based global environmental standards that define performance levels to be adhered to by all of the firm's facilities around the world, even where this entails going beyond local regulatory requirements. These firm-based standards are themselves becoming an important platform for firm-based learning, where solutions to environmental challenges identified within one part of the firm are shared throughout the enterprise (Angel and Rock 2005).

Thirdly, the greening of industry raises important questions regarding flows of capital, technology and information that are at the heart of current debates around economic globalization. One of the most crucial issues with respect to environmental and resource conditions in rapidly industrializing and urbanizing economies is the extent to which the scale effects of industrial and urban growth can be offset by improvements in energy, materials and resource efficiency. With new capital investment comes the opportunity to take advantage of cutting-edge technologies that are potentially far more energy and resource efficient than older capital stock. With much of the technology and capital equipment being deployed in developing economies sourced from OECD countries, this raises fundamental questions regarding the diffusion and adoption of technology on a global scale, and the way in which such technology transfer is structured by processes of foreign direct investment and supply chain linkages. In addition, patterns of technology transfer, adoption and use typically involve intersections between international flows of capital and technology, and local, regional and national conditions that influence the rate and effectiveness of technology adoption. These classic 'global–local' intersections are at the core of economic geography's analysis of the dynamics of regional economic development.

For all the opportunities for economic geography to engage with the greening of industry, the approach described above does not address fully the significance of an engagement with issues of environment and resources for theory development in economic geography. In most cases research within this framework takes existing economic geography theory and applies this work to the domain of the greening of industry. In addition, emerging research into the greening of industry tends to be focused on individual firms and industries, and does not engage in the type of structural analysis that has been an important part of theory development in economic geography over the past three decades. But perhaps most fundamentally, work within the greening of industry framework does not engage fully with the material basis of economic activity. The environment is examined as inputs and outputs to economic change rather than constituting and theorizing these processes of economic change as simultaneously material and social in form.

To go beyond the greening of industry is to be engaged in what might be called a political ecology of industrial change. Then the question becomes how to theorize industrial change as a process that is as much about flows of materials and resources as it is about flows of capital, technology, products and services. This is a field of enquiry that is in its infancy within geography and other disciplines (although arguably ecological economics has gone someway to developing methodologies for assessing economic change in these terms). There are many broad questions of interest. For example, to what degree are the current 'worlds of production' and attendant geographies of economic activity predicated upon existing resource and material foundations, or more generally, upon currently constituted human-environment relations? Would a substantial shift toward renewable sources of energy, away from the use of certain toxic chemicals, toward less water-intensive production processes, or any number of other changes in the material foundation of economies be of significance for the organization and geography of industrial activity – beyond the price effects that are part of the current calculus of economic analysis? This is in part the question that ecological modernization is asking with respect to institutions and governance in capitalist societies (the capacity to achieve dramatically different human environment relations in the context of existing dominant capitalist social relations).

Earlier in this chapter two axes of geographical scholarship were described, one defined in terms of studies of human-environment relations and the other in terms of spatiality of economic and social processes. An environmental economic geography has the potential to make important contributions at the intersection of these two axes of enquiry. Within an environmental economic geography, economic change and attendant livelihood and development effects remain of central concern. But these dynamics of economic change are theorized as both material and social processes, and are measured in terms of their impacts on environmental quality and resource footprints, as well as jobs created, products sold and profits made. Certainly these issues are of compelling public concern, and for many firms and industries, they are an increasingly important part of the day-to-day management of industrial activity.

Note

1. RoHS is the European Union directive on the restriction of the use of certain hazardous substances in electrical and electronic equipment.

References

Angel, D. (2000) 'Environmental innovation and regulation', in G. Clark, M. Gertler and M. Feldman (eds) *Handbook of Economic Geography*, pp. 607–22. Oxford: Oxford University Press.

Angel, D. and M. Rock (2005) 'Global standards and the environmental performance of industry', *Environment and Planning A*, 37: 1903–18.

Bakker, K. (2004) *An Uncooperative Commodity: Privatizing Water in England and Wales*. Oxford: Oxford University Press.

Braun, B. (2005) 'Environmental issues: writing a more-than-human urban geography', *Progress in Human Geography*, 29: 635–50.

Bridge, G. (2002) 'Grounding globalization: the prospects and perils of linking economic processes of globalization to environmental outcomes', *Economic Geography*, 78: 361–86.

Castree, N. (2004) 'Environmental issues: signals in the noise', *Progress in Human Geography*, 28: 79–90.

Castree, N. and B. Braun (2001) *Social Nature: Theory, Practice and Politics*. New York: Blackwell.

Clark, G., M. Gertler and M. Feldman (eds) (2000) *Handbook of Economic Geography*. Oxford: Oxford University Press.

Dicken, P. (1986) *Global Shift*, 1st edn. London: Harper and Row.

Fitzsimmons, M. (1989) 'The matter of nature', *Antipode*, 21: 106–20.

Gibbs, D. (2002) *Local Economic Development and the Environment*. London: Routledge.

Gibbs, D. (2003) 'Ecological modernisation and local economic development: the growth of eco-industrial development initiatives', *International Journal of Environment and Sustainable Development*, 2: 1–17.

Marsden, T. K., G. Bridge and P. McManis (2003) 'Guest editorial: the next new thing? Biotechnology and its discontents', *Geoforum*, 34: 165–75.
Murdoch, J., T. Marsden and J. Banks (2000) 'Quality, nature and embeddedness: some theoretical considerations in the context of the food sector', *Economic Geography*, 76: 107–25.
O'Brien, K. and R. Leichenko (2003) 'Winners and losers in the context of global change', *Annals of the Association of American Geographers*, 93: 89–103.
Peet, R and M. Watts (1996) *Liberation Ecologies: Environment, Development, Social Movements*. London: Routledge.
Robbins, P. and J. Sharp (2003) 'Producing and consuming chemicals: the moral economy of the American lawn', *Economic Geography*, 79: 425–51.
Rock, M. and D. Angel (2005) *Industrial Transformation in the Developing World*. Oxford: Oxford University Press.
Scott, A. (2000) 'Economic geography: the great half-century', in G. Clark, M. Gertler, M. Feldman (eds) *Handbook of Economic Geography*, pp. 18–44. Oxford: Oxford University Press.
Turner, B. L. (2002) 'Contested identities: human-environment geography and disciplinary implications in a restructuring academy', *Annals of the Association of American Geographers*, 92: 52–74.
World Bank (1999) *Greening Industry*. Washington, DC: World Bank.

11 Digitizing services

What stays where and why

Martin Kenney and Rafiq Dossani

The spatial extension and deepening of capitalism has been a topic of interest to geographers, other social scientists, and activists since, at least, Lenin. This topic has reappeared on the public agenda recently under the rubric of 'globalization'. Once again, the spatial redistribution of economic activities is sparking enormous controversy and opinions from nearly every philosophical position. This chapter considers two dimensions of this enormous topic and argues that neither dimension has received sufficient attention from geographers. The first dimension is the role of technological advancement in transportation and communication technologies in a capitalist system. The second dimension is the development of a global division of labor in service provision.

In 1980, Frobel et al. hypothesized that a new international division of labor was being created within which low skilled manufacturing work, which had previously been located in the developed nations, was being transferred to developing nations to take advantage of low-waged, mostly female workers. At the time, they suggested that this was an inherently unequal exchange and that the workers in both locations were victims of this relocation. This essay will not engage the debate about the exploitation of low-wage workers in developing nations except to assert that the plight of these workers has received an enormous amount of attention from geographers, sociologists, anthropologists, professors of women's studies, and social activists. Quite naturally, in their zeal to struggle against the very real and shocking work conditions under which these workers labor, they have focused on a few industries particularly garments and shoes (industries known in the developed world for shocking labor conditions), and, to a much lesser extent, electronics. It is remarkable how social science researchers have reduced the integration of the developing nations into the global economy to garments and shoes. This fixation has had the unfortunate effect of resulting in a one-sided understanding of globalization

This essay directs attention away from these infernal mills to the two dimensions that I believe will have a far more significant effect. Consider the implications of how the rapidly evolving global transportation and communications infrastructure is tying the global economy more firmly together. The globalization of manufacturing is being followed by a global redistribution of white-collar work. This has only recently begun. As this advances, it will lead to a fundamental geographic redistribution of work that is also nearly certain to have profound effects on the global economy. These two themes are not new as Dicken (2004) touched upon these in his lament that geography was being left out of the globalization discussion. To presage my concluding discussion I will argue that geography has been so swept into the study of clusters and the interest in cultural studies that it is missing the macroforces that are transforming the world economy.

This chapter speculates on the implications of the digitization of work and what the global improvement in telecommunications and transportation networks means for the creation of a global work force and, by extension, a global labor market. This will threaten those in developed nations whose skill levels are not sufficiently superior to those in developing nations to justify receiving developed nation's wages. For all economies it suggests, *ceteris paribus*, that workers wherever they are will be rewarded more equally.

Transportation and communication systems

Among the classical social theorists, it was Marx who was the most fascinated by transportation and communication systems for the development of capitalism. He considered them most directly in volume two of *Capital* where he recognized their centrality in reducing turnover time by increasing the velocity of all types of capital. These technologies form the arteries by which economic actors are interconnected. Their importance in terms of enabling both greater speed and greater throughput has been recognized by many (Chandler 1977; Fishlow 1965). When a new medium of communication or transportation emerges or a previous media is dramatically transformed, such as recently occurred with the transformation of telecommunication networks from analog to digital transmission, opportunities are created for entrepreneurs to utilize the new media for both organizational and spatial transformations (Fields 2004).

Transportation

In *The Wealth of Nations* Adam Smith recognized the importance of transportation for the development of markets. And yet, until the application of the steam engine to locomotion, for all intents and purposes, transportation was limited to either animate power or the speed provided by natural forces such as winds and currents. The steam engine began revolutions in transportation, both on land and at sea, and then eventually in the air that permitted the relocation of production sites, e.g. internal combustion engines dramatically increased the power available for locomotion and speeding the movement of goods and people.

These innovations are in fact sociotechnical. For example, the global maritime infrastructure was the result of a long coevolution of networks including shipping firms, freight forwarders, ports, insurers, brokers, and many other intermediaries.

Transportation technology has also affected organizational forms. Braudel (1982: 371) reminds us that in the 1400s trade was so risky and difficult that most merchants fitted out their own ships because 'the risks and the cost price relative to the cargoes transported were so great in long-distance shipping that they made transporting as a simple freight industry virtually unthinkable'. In other words, the merchant had to integrate ship ownership and operation – there was no opportunity for a division of labor. Later, the development of the shipping networks would coevolve with the increase of trade and the further development of international finance (Miller 2003: 4).

For today's industries, the most important transportation technologies are air freight, which has grown remarkably during the last two decades, and shipping containerization, which has sped surface transport and lowered its costs and risks. Containerization is at the core of intermodalism, for example, the ability to move cargo in the same containers by sea or land. With the standardization of the shipping container's dimensions, logistical planning was simplified. As the stevedore was replaced by crane operator loading and unloading 20- and 40-foot cargo containers from specialized container ships, the delays at the ports were drastically reduced.

The intermodal shipping container provided a base for further innovations that would have significant geographical implications by reducing the friction of geography. For example, United States retailers monitor their sales in real-time permitting them to reorder goods electronically shortening lead times. The orders are not only transmitted, to say China, but they also inform the vendor loading the order. The United States delivery route for the container has been established prior to loading eliminating the need to send the products to a warehouse for storage or sorting. The container is lifted off the ship directly onto a truck that then delivers directly to a store.

The shipping container has become the critical package in world trade, and container traffic, along with air cargo, is the fundamental measure for the growth of trade. The emergence of China as the global workshop can be seen in the rapid growth in the number of containers moving through its ports. In 2003, the container throughput of Chinese ports reached 48 million ton equivalent units (TEUs), the largest number of containers traversing any nation in the world. From January through September 2004, container throughput was 43.7 million TEUs, a 27.2 percent increase over the same period in 2003. Hong Kong, which serves South China, is already the busiest port in the world, but Shanghai area ports are rapidly gaining on it.

Air transport was the other key transportation system and it has been growing rapidly also. Though the bulk of the finished good flows through the medium of shipping containers, for the highest value-added items that are most subject to decay or obsolescence air transport is the method of choice as it is for people traveling long distances. For semiconductor chips, hard disk drives, fresh fish, and many other products that lose their value rapidly airfreight has become critical. Elaborate service infrastructures have developed to ensure that goods air transported are not delayed in their movement (Leinbach and Bowen 2004).

In terms of costs, air transport has been decreasing at a rate of 3 percent per annum. For locations wishing to ascend the value-added hierarchy an international airport is a critical infrastructural requirement, whether one is exporting cut flowers from Bogota to Miami or newly packaged integrated circuits from Penang, Malaysia. The 'fastest growing market of all for air freight is in IT goods from Asia to Europe and North America, [and this represents] 40% of the total shipments by tonnage and nearly 75% by value' (Butterworth-Hayes 2005). To be a global-class industrial center, global quality infrastructure has become a requirement.

Telecommunications

The effect of communications on geography has been dramatic. For example, Febvre and Martin (1976) titled an entire chapter in their treatise on the innovation of the moveable-type printed book as *The Book as a Force for Change*. The printed book increased the volume transmitted and, by lowering the cost of reproduction, expanded the number of persons capable of accessing the knowledge and information. But, more important, the accelerated circulation of information, in the form of codified knowledge, sped the creation of new knowledge and information, forming a virtuous circle of knowledge growth that continues to this day. To illustrate, typeset books circulated the heretical views of Galileo and Copernicus far more rapidly than hand copies of an original manuscript ever could have.

Twenty-five years ago, telecommunications capacity was concentrated in the developed nations. Phone calls to India, China, or even Mexico were expensive and the quality of service was low or even extremely low. In the 1980s this began to change as telecommunications was deregulated and there was increasing pressure for improved and lower cost service. With the construction of new fiber optics undersea cables during the Internet Bubble of the 1990s, a dynamic of double-digit percentage price declines per annum for international service was set in motion.

India is an excellent example of a formerly bandwidth poor nation whose telecommunications infrastructure has improved dramatically. To illustrate, India's international submarine cable capacity grew from 31 gigabytes per second (Gbps) in 2001 to 541 Gbps by the end of 2004. In China, the cost of international service has plummeted to the point at which it was possible to buy telephone cards offering United States to China calling for about $.03 per minute. The final chapter is the coming voice-over-Internet-Protocol telephony that will create always-on connections priced at a low monthly fee. With this the cost of transferring information long distance will no longer be significant. In this respect the prediction 'death of distance' is being fulfilled (Cairncross 1997).

Telecommunications linkages, capacity, and cost will no longer be a significant differentiator in providing workers protection from competition. Moreover, the rapidity with which service can be provided means that any part of the world having customers willing to pay for bandwidth or workers who can be profitably integrated into the global economy will receive service. Of course, the telecommunications networks transmit data and information, they do not create knowledge nor can they easily transmit tacit knowledge. Even seeing a person during a teleconference is not the same as in-person interaction, which provides the multi-channeled analog information coming from the person and the context. What the telecommunications networks do provide is an increased ability to share explicit knowledge and information.

The changing organizational and technical aspects of transportation and communication networks have provided enormous impetus to the expansion of the economy especially in terms of integrating labor into the global economy. The impacts of this expansion upon manufacturing workers is already well known. However, the most recent changes in the telecommunications infrastructure is now threatening to have a similar impact on service work, not only at the low end in tasks such as data entry, but even more interestingly on high-end service and research and development (R&D) jobs. The implications of this are examined in the next section.

The changing global geography of service work[1]

The quickening pace of relocation of services overseas, calls into question our normal thinking about what a 'service' is.[2] The ongoing transformation of the global location of service delivery is captured in Jones (2005) where he discusses how service firms are trying to transform their operations from national to functional operations in which there are global competency centers for particular functions. In the case of manufacturing, most persons would agree that it is a process that involves the transformation of a tangible good, though as Sturgeon (2002) has pointed out there is an entire industry providing manufacturing as a 'service'. In terms of the location and discharge of manufacturing, it is generally accepted that manufacturing does not necessarily require constant face-to-face contact between the producer and consumer. Manufacturing usually creates a good that can be stored, thereby allowing a physical separation of the buyer and the seller.

Services have usually been defined as the opposite of manufacturing: they are transactions of intangible, nonstorable goods, requiring that client and vendor be face-to-face while the service was being delivered. For example, Gadrey and Gallouj (1998) define services as goods that are 'intangible, cosubstantial (e.g. they cannot be held in stock) and coproduced (e.g. very often their production/consumption requires the cooperation between users and producers).' This is obviously true when the service requires face-to-face experience, such as receiving a haircut, but also true when the 'service experience' did not require proximity, such as when a bank's client wants to check their bank balance.

These definitions, though never exhaustive in prior periods, are now under great stress in the digital age that was inaugurated by the application of von Neumann's principles and actualized by the development of low cost techniques for information digitization, transmission, and processing. These new technologies have had a profound impact on the discharge of services. Paraphrasing futurists of the 1960s, if robots were going to change the factory of the future, digitization would change the office of the future. First, the digital age allowed (or, at least, dramatically eased) the conversion of service flows into stocks of information, making it possible to store (or, more properly, productize) a service. For example, a legal opinion that earlier had to be delivered to the client in person could now be prepared as a computer document and transmitted to the client over email or, better yet, encoded into software. Easy storage and transmission allowed for the physical separation of the client and vendor as well as their separation in time. It also facilitated the separation of services into components that were standardized and could be prepared in advance (such as a template for a legal opinion) and other components that were customized for the client (such as the opinion itself) or remained nonstorable. Taking advantage of the possibility of subdividing tasks and the economies that come with a division of labor, this reduced costs by offering the possibility of preparing the standardized components with lower cost labor and, possibly, at another location or if all the necessary materials were digitized then the entire product could be produced at another location.

The second fundamental impact was the conversion of non-information service flows into information service flows. For example, sampling of tangible goods by a buyer visiting a showroom is increasingly being replaced by virtual samples delivered over the Internet. Once converted to an information flow, the service may also then be converted into a stock of information, as noted earlier, and subjected to the above mentioned forces of cost reduction through standardization and remote production.

By enabling transmission and storability, the digital age accelerated the relocation of services. The offshoring of services such as writing software was enabled by digitized storage and facilitated by the adoption of standardized programming languages. As transmission costs fell (just as digital storage costs had earlier fallen), even non-storable services, such as customer care, could be relocated. As a result, any location with the requisite labor power could become a services producer. The range of such services is massive, and includes back office services such as payroll, front-line services such as customer care and telemedicine, patent preparation, equity analysis, medical transcription, medical imaging interpretation, remote facilities management, and, of course, software services such as programming and remote IT infrastructure management.

The current emerging insertion of India into the global economy illustrates how activities that were once considered planted in the developed world are being uprooted and redistributed globally (Dossani and Kenney 2003). The thesis is that the tasks being moved are not only the simple commoditized activities that most persons suspect will be relocated, but rather there are a number of high-value activities also being transferred, and that it is these that give us a far better insight into the future geography of innovation and the location of value creation.

India's entry into the global economy came through the very simple stratagem initiated by United States computer firm, Burroughs Corporation, which suggested to its Indian affiliate, Tata Consultancy Services, that it transfer some Indian software engineers to the United States to install software. This was the genesis of 'body shopping' (Heeks 1996). Gradually, these Indian firms shifted to an offshore model where they would do the coding in India. Roughly contemporaneously in 1986 a group of American multi-national corporations (MNCs) led by Texas Instruments began doing software development work in Bangalore after the government guaranteed them satellite bandwidth. Very soon, these MNCs discovered that Indians were extremely capable particularly in areas like algorithm development. In the 1990s, some of the world's largest financial institutions such as Citicorp, American Express, and General Electric Capital Corporation also established software development operations in India. General Electric and a number of other major medical imaging firms located sales, marketing, then production, and finally R&D operations in India because of the large market for fetal imaging. For a number of reasons, MNCs became increasingly acquainted with the capabilities of Indian workers.

The 1990s were a tumultuous period in American capitalism as it experienced the largest stock market bubble since the 1920s. The core of the bubble was technology, communications, and, most centrally, the Internet. Given the massive databases, website development, and other chores that came with the feverish panic to create an online presence, there was a belief that the developed nations were running out of software programmers. To remedy these perceived shortages, foreign programmers were welcomed into the United States even as firms became increasingly willing to offshore development work to lower-cost environments.

Increasingly, the objects of white-collar work, for example, papers, data, files, and images, were digitized or could be scanned and made digital. Though not immediately obvious, what this meant is that the information within these items was being dematerialized. Even as existing information was digitized, there was a proliferation of sensors, processors etc. that were creating an even larger sea of information to be processed and interpreted. Finally, this information could, in principle, be transferred to any location having two wires – one for electricity and the other carrying communications.

White-collar work was increasingly undertaken on digitized images on a screen. Previously a business process such as filing, researching and adjudicating an insurance claim triggered a set of actions that moved pieces of paper from one office to the next downstream to final resolution; very often generating yet more paper as it moved. Moving these papers a long distance was almost impossible in terms of prohibitive costs, risks of misplacement, and delays. These barriers entirely disappeared once the information was digitized; now the information could flow at the speed of electrons.

How deep the offshoring process will be is inherently unknowable. Consider the promise of telemedicine. If, through the use of cameras and telecommunications linkages, a doctor in an urban medical center can remotely diagnose a person in a rural American farm community, then the doctor can just as easily be in New Delhi or Buenos Aires. For example, in diagnostic endoscopy the doctor uses a digital image for guidance, why does the doctor have to be located in the surgery? Why not in a room across the street or anywhere else in the world equipped with high-speed Internet access? Often, these technical possibilities must be coupled with social innovations. These new technologies do not presage the replacement of all doctors, but, more probably, a reengineering of what the spatial and hierarchical division of labor should be. Already, x-ray and other medical images are interpreted in India for the American market. This illustration suggests that the geographical division of labor of this and many other production processes is likely to become increasingly complex.

The medical example suggests something even more interesting, namely that well-paying occupations in which high levels of discretion and skill are required, and thus have normally been considered immune to global competition may, at least, partially be in the process of becoming vulnerable to relocation. Geographers interested in labor process questions could provide important insight into how this will develop.

Discussion

Capitalism had an uncanny ability to dissolve and reorganize previous arrangements. One of its most powerful levers for melting ossified social arrangements was the application of science and technology to the workplace. Capitalism, through the medium of the entrepreneur, constantly searches for opportunities to integrate more people into its orbit, and today we are seeing the integration of both India and China. At this point, it is impossible to predict what the outcome of this process will be, but there is ample evidence that it will be profound.

Economic geography can play a central role in providing a better understanding of globalization and its implications for the global economy. There is a continuing need for theoretically informed study of globalization. There is a paucity of studies on the telecommunications and transportation infrastructures that are facilitating globalization. The remarkable emergence of China onto the global economy has received only minimal attention from geographers, this should be a very profitable vein of research and Chinese scholars are eager to cooperate with those in the West. The rise of China is already obvious, if understudied, the case of India is almost entirely unexamined, and the Indian case is probably more important, because it immediately leads one into a contemplation of what is the nature of services, or, what could be termed, 'mental' labor. Trying to better comprehend the redistribution of this mental labor globally is possibly the most interesting and possibly most profound new wrinkle in the continuing evolution of the global economy.

With such exciting topics, the growing awareness in all of the social sciences of the importance of the spatial, the interest in understanding globalization, and the intellectual ferment within economic geography provides ample grounds for optimism for the future of economic geography. To justify this optimism, economic geography must escape from the cul de sac of post-modernist (best left in architecture where it made sense), deconstructionist (best left in literature) cultural studies to reengage with studies of the real world of economic action, otherwise it seems likely that the core topics of economic geography will be absorbed by the other social sciences and as Dicken (2004) so well sums it up, 'geography will miss the boat'.

Notes

1. This section draws heavily upon Dossani and Kenney (2003).
2. On the geography of producer services in the US, see, for example, Beyers and Lindahl (1996), though producer services is too narrow as a description of what is being relocated offshore.

References

Beyers, W. B. and D. P. Lindahl (1996) 'Strategic behavior and development sequences in the producer service businesses', *Environment and Planning A*, 29: 887–912.

Braudel, F. (1982) *The Wheels of Commerce.* New York: Harper & Rowe Publishers.

Butterworth-Hayes, P. (2005) 'The changing shape of the air cargo market', *Aerospace America*, 8–10 March.

Cairncross, F. (1997) *The Death of Distance: How The Communications Revolution will Change our Lives.* Boston: Harvard Business School Press.

Chandler, A. D. Jr (1977) *The Visible Hand: The Managerial Revolution in American Business.* Cambridge: Harvard University Press.

Dicken, P. (2004) 'Geographers and "globalization": (yet) another missed boat?', *Transactions of the Institute of British Geographers*, 29(1): 5–26.

Dossani, R. and M. Kenney (2003) 'Lift and shift: moving the back office to India', *Information Technology and International Development*, 1(2): 21–37.

Febvre, L. and H.-J. Martin (1976) *The Coming of the Book.* London: Verso.

Fields, G. (2004) *Territories of Profit: Communications, Capitalist Development, and the Innovative Enterprises of G. F. Swift and Dell Computer.* Stanford: Stanford University Press.

Fishlow, A. (1965) *American Railroads and the Transformation of the Ante-Bellum Economy.* Cambridge: Harvard University Press.

Frobel, F., J. Heinrichs, and O. Kreye (1980) *The New International Division of Labour: Structural Unemployment in Industrialised Countries and Industrialisation in Developing Countries.* Cambridge: Cambridge University Press.

Gadrey, J. and F. Gallouj (1998) 'The provider-customer interface in business and professional services', *The Services Industries Journal*, 18(2): 1–15.

Heeks, R. (1996) *India's Software Industry.* Thousand Oaks, CA and London: Sage Publications.

Jones, A. (2005) 'Truly global corporations: theorizing "organizational globalization" in advanced business services', *Journal of Economic Geography*, 5: 177–200.

Leinbach, T. R. and J. T. Bowen Jr (2004) 'Air cargo services and the electronics industry in southeast Asia', *Journal of Economic Geography*, 4: 299–321.

Miller, M. B. (2003) 'The business trip: maritime networks in the twentieth century', *Business History Review*, 77(Spring): 1–32.

Sturgeon, T. J. (2002) 'Modular production networks: a new American model of industrial organization', *Industrial and Corporate Change*, 11(3): 451–96.

12 Globalizing Asian capitalisms

An economic-geographical perspective

Henry Wai-Chung Yeung

Introducing the contradiction

By the late 1960s and throughout the 1970s, there was no doubt that American firms and their counterparts from Western Europe were spearheading the globalization of economic activities through their cross-border foreign direct investments (FDI). This special genre of the capitalist firm has been described as the modern transnational corporation (TNC). Its rapid emergence especially since the 1960s became the cause of some serious political and economic concerns by the early 1970s (Vernon 1971). In economic geography, much attention during the same period was paid to the role of branch plants externally controlled by these TNCs in local and regional economic development in advanced industrialized economies. The analytical foci were placed on industrial linkages and decision-making in these branch plants and their local and regional economic impact (Dicken 1976; Hamilton 1974).

Amidst this growing fear of external control by TNCs of local and regional economies in the host developed and developing countries, a parallel, but slightly belated, process of outward investment by business firms based in what was then known as the 'Third World' began to gather momentum (Agmon and Kindleberger 1977; Lall 1983; Wells 1983). By the late 1970s, the rise of these so-called 'Third World multinationals' was hailed by two business school professors in their *Harvard Business Review* article as 'only yesterday an apparent contradiction in terms' and 'now a serious force in the development process' (Heenan and Keegan 1979: 109). They further argued that: 'The multinational corporation, long regarded by its opponents as the unique instrument of capitalist oppression against the impoverished world, could prove to be the tool by which the impoverished world builds prosperity'.

What does this 'contradiction' of TNCs from developing countries entail and how does it contribute to the globalization and transformation of their home capitalist economies? In this chapter, I address this research problem in relation to the distinctive contributions made possible by adopting an economic-geographical perspective. In particular, I argue that an explicit attention to *business and production networks* spanning different spaces and scales has enabled economic geographers to draw significant interconnections between the rapid emergence of these developing country TNCs and the tremendous transformations in their home economies in Asia during the past three decades – a phenomenon broadly known as 'globalizing Asian capitalisms'. Indeed, many TNCs from developing Asian countries had a humble origin as regional trading and commercial ventures; they had internationalized across national boundaries as early as the late nineteenth century. Their participation in globalization, however, did not occur until much later in the 1980s when the global economy was increasingly competitive, their organizational capabilities were much more consolidated, and their home governments were serious about growing 'national champions' (see Yeung 1999). In this sense, these Asian-origin TNCs are important *conduits* through which their burgeoning domestic economies become articulated into the global economy.

From an anomaly to major leagues: Asian firms in the global economy

By the early 1990s, many Asian TNCs had already grown out of their former shadow as an anomaly in the global economy. Some of them had become significant competitors in such diverse industries as air and sea transportation (e.g. Singapore Airlines and Evergreen Shipping Lines), consumer and computer electronics (e.g. Samsung and Acer), semiconductors (e.g. TSMC and UMC), textile and garments (e.g. Fountain Set and Esquel Group), hotels (e.g. Shangri-La Hotels), property development and construction (e.g. Hutchison Whampoa), and so on. They were no longer mere followers of giant TNCs from advanced industrialized economies; they had become in their own right major competitors in the global economy. To account for this significant transformation in the organizational and competitive dynamics of Asian TNCs, economic geographers naturally turned to leading theoretical perspectives in international economics and international business studies such as Stephen Hymer's market power perspective, Raymond Vernon's product life-cycle hypothesis, Peter Buckley and Mark Casson's internalization theory, and John Dunning's eclectic framework of international production. While these perspectives explain reasonably well why domestic firms engage in international production from an economic standpoint, they fail to account for the processes of *organizing* international business activities. Moreover, they are overtly based on economic considerations, ignoring the peculiar social and institutional contexts from which these Asian TNCs emerge. One such important Asian context is the role of *personal and social networks* in engendering economic processes; this critical context is clearly difficult, if not impossible, to be subsumed under the conceptual apparatus of transaction cost economics that tends to dominate most of these economic models of TNCs.

At around the same time, a heated debate in economic geography was focusing on the changing organization of production systems in different industrial districts and new industrial spaces, again, in advanced industrialized economies – the so-called flexible specialization debate. Back then, this debate represented a telling move away from the Marxist structural understanding of the role of firms and their production systems in contemporary capitalism that came to dominate economic geography of the 1970s and the 1980s. Instead, the flexible specialization debate focused on how business firms are actively transforming their organizational processes in order to ride out of capitalist crises. In initiating what was subsequently termed the 'cultural turn' in economic geography, Peter Dicken and Nigel Thrift (1992) were one of the earliest in economic geography to incorporate the concept *embeddedness* in their conceptualization of the dynamic organization of business firms (Yeung 2003). Originally developed by Karl Polanyi to explain capitalism's institutional transformation, the concept was reinvigorated by economic sociologist Mark Granovetter (1985) to counter what he called 'under-socialized' account of economic life found in most transaction cost analysis of firms and markets. To Dicken and Thrift (1992: 285–6; original italics), the organization of production chains and production systems through business organizations can thus be conceptualized 'as a complex set of *networks* of interrelationships between firms which have differing degrees of *power and influence*'.

Inspired by Dicken and Thrift's (1992) reconceptualization, economic geographers have subsequently expanded on this notion of social embeddedness of business firms in a business network perspective that accounts for both economic and non-economic relations at the intra-firm, inter-firm, and extra-firm dimensions (Grabher 1993, 2006; Yeung 1994). This network perspective extends earlier work in organizational studies that focuses primarily on inter-organizational networks. The advantage of this network perspective rests in its capacity to incorporate different loci of power and control within specific firms (intra-firm dimension) and significant bargaining and cooperative relationships between business firms and other institutions such as government agencies and NGOs (extra-firm dimension). Unknowingly then, this network perspective has anticipated the recent 'relational turn' in economic geography through which we place great analytical emphasis on dynamic relations among social actors in producing diverse economic geographies (see Yeung 2005).

More specifically, economic geographers have successfully applied this business network perspective and explained the internationalization of entrepreneurs and firms from Asia to North America (Hsu and Saxenian 2000; Mitchell 1995; Olds 2001; Zhou and Tseng 2001) and within Asia (Hsing 1998; Leung 1993; Qiu 2005; Yang and Hsia 2006; Yeung 1997, 1998). This economic-geographical approach pays special attention to different forms of network relationships in which Asian entrepreneurs and their firms are embedded. In doing so, it has effectively transcended the economistic approach to explaining TNCs and FDI commonly found in earlier studies of 'Third World multinationals' (e.g. Lall 1983; Wells 1983; cf. Yeung 1999). This strength of the network perspective on TNC activities is significant as most studies of TNCs and FDI throughout the 1980s and early 1990s have focused narrowly on transaction cost economizing strategies of these capitalist firms. The network perspective pursued by economic geographers thus allows us to link economic outcomes of TNC activities in one place to non-economic strategies pursued by social actors (e.g. entrepreneurs and managers) elsewhere. It represents one of the most unique economic-geographical insights – our attentive consideration of *economic relations spanning different spaces and places*. Instead of seeing economic action as discrete decisions in the context of arm's length market transactions – a common problem in most economic models of TNCs, the network perspective identifies relational synergistic effects embedded in ongoing interaction among social actors.

One such critical synergy in the case of large Asian firms and TNCs is the role of *family* in their ownership and control. The *World Investment Report 2004* describes that in 2002, some 32 of the world's top 50 TNCs from developing economies ranked by foreign assets originated from Asia. At least 18 of these 32 Asian TNCs are family-owned and controlled. The rest are mostly state-owned TNCs from Singapore and Malaysia (about 10 of them). The dominant role of family in Asian business and thus the emergence of Asian TNCs may not be surprising (La Porta et al. 1999). What is rather unexpected is the rapid growth and expansion of these large TNCs from Asia to become significant competitors in the global economy by the turn of the new millennium. A direct research problem emanating from this economic-geographical phenomenon is the concomitant evolution of family business and homegrown TNCs in Asia.

Globalizing Asian capitalism: dynamics of change and adjustments

In explaining the cross-border investment activity of leading Asian firms and uncovering their closely-knit network relationships, economic geographers have realized that institutionalized patterns of political, economic, and cultural processes in their home economy must have played a decisive role much more than simply as a contextual factor. Rather, these institutionalized structures must have a direct bearing on the processes and outcomes of the globalization of Asian firms. For example, despite their very similar historical and geographical contexts, TNCs from Hong Kong and Singapore exhibit very different ownership structures, corporate governance, and entrepreneurial strategies (Yeung 2002). Whereas large Hong Kong TNCs tend to be organized in the form of entrepreneurial family-controlled enterprises (Leung 1993; Mitchell 1995; Olds 2001), large Singapore-based TNCs are mostly state-owned and managed by former civil servants. TNCs from both economies also have contrasting business practices and financial discipline. The same economic-geographical observation can also be applied to TNCs from Taiwan and South Korea, both of which are former Japanese colonies and yet have produced TNCs of very different organizational structures and industrial specialization (Cho 1997; Hsing 1998; Hsu and Saxenian 2001; Park 1996). These enduring home-economy institutional imprints on national firms and their globalization strategies might seem rather odd at a time (late 1990s) when the popular debate on globalization had moved in favour of a wholesale convergence in global business norms and practices in an allegedly 'borderless world' – a derisive term coined by business guru Kenichi Ohmae (1990).

In contesting this global convergence perspective, organizational sociologist Richard Whitley (1992) has developed a business system approach to understanding the enduring influence of home country on business firms. In this approach, Whitley (1992) focuses on enduring institutions in national economies such as political systems, economic beliefs, cultural norms, employment relations, and so on. He argues that business systems vary significantly across different capitalist economies. As different forms of capitalism entail different business systems, Whitley (1999) proposes the idea of 'divergent capitalisms' to describe the social structuring of business systems. Joining a number of other prominent political scientists and economic sociologists, Whitley (1992, 1999) subscribes to the view that globalization continues to accentuate systemic differences that exist prior to the onslaught of global processes. This is the so-called 'divergence school' in studies of global political economy.

Confronting with these two contrasting schools of thought on economic globalization and its impact on capitalist economies, economic geographers have found supporting evidence for both schools in their studies of the political-economic structures of many Asian economies dominated by ethnic Chinese business firms (e.g. Indonesia, Hong Kong, Malaysia, Singapore, Taiwan, and Thailand). Just when leading Asian firms are becoming significant players in the competitive global economy through their globalization activities, their home economies are exerting structural influences on their business strategies and corporate practices. To reconcile this apparent paradox in understanding Asian capitalism, some economic geographers have begun to take on an actor-specific approach to understand the dynamics of Asian capitalism and to transcend the dualistic thinking manifested in much of the convergence-divergence debate (Olds 2001; Olds and Yeung 1999; Yeung 2000, 2004).

Drawing upon Nigel Thrift's (1996) geographical adaptation of actor-network theory, these economic geographers have revisited the earlier business network perspective and incorporated heterogeneous relationships among actors as a central driving force in their network analysis (see Coe et al. 2004; Dicken et al. 2001; Henderson et al. 2002). This actor-specific approach moves away from a structural reading of the global economy in which enduring institutions of capitalism are seen as either outdated models to be replaced by a more superior form of global economic coordination (the convergence school) or persistent stumbling blocks to global convergence (the divergence school). Instead, economic geographers have placed significant analytical emphasis on key capitalist actors such as business firms and examined how their participation in globalization processes could have generated powerful 'bottom-up' effects to transform those enduring home economy structures and institutions. In the case of Asian capitalism, this actor-specific approach has worked well when we consider the dynamics of ethnic Chinese business firms in engaging with globalization tendencies and in bringing about significant transformations in their home economies.

One of the most interesting geographical insights from this actor-specific approach to globalizing Asian capitalism is the analytical role of spatial scales in helping us to draw connections across drastically different processes. Take culture as an example. The Asian Confucian heritage and culture came to the forefront of debate on the meteoric rise of East and Southeast Asia during the late 1980s and the early 1990s. Two significant publications by sociologists Peter Berger and Michael Hsiao (1988) and Gordon Redding (1990) have established firmly the 'culturalist explanation' of capitalist economic development in Asia. To put it in brief, the Confucian ethic of hard work and the cooperative relations in a family-centric form of economic organization have allegedly made it possible for these Asian economies to experience rapid industrialization and economic development. Rapid economic development has occurred in many Asian economies, despite the existence of authoritarian states in most of them. This culturalist-inspired explanation of the Asian 'miracle economies' not only fits well into World Bank's (1993) own triumphant assessment, but also perpetuates the myth that these cultural imprints of work ethics and family values are unique to Asian business systems – a gross generalization reinforced strongly in Whitley's (1992) approach.

The key problem in this argument for Asian exceptionalism is its serious confusion of different spatial scales of representation. The critical question is not so much about the existence of national cultures, but rather their *spatial boundedness*. While it perhaps makes good sense to talk about how economic life is bounded by culture in traditional societies relatively closed to the outside world, globalization – be it political, economic, and cultural – has clearly diminished the possibility of using the *national* scale as a bounded space to analyse economic processes, let alone their dynamic and transformative nature. This is particularly so if these economic processes are transcending and transversing different spaces that range from global and regional to local economies. The culturalist argument for Asian exceptionalism thus falls short of accounting for dynamic transformations brought about by globalization tendencies precisely because of its 'over-socialized' analytical framing, to borrow from Granovetter (1985), and its 'under-spatialized' worldview.

By adopting differential spatial scales in our analysis of Asian capitalism, economic geographers are able to connect changes at the level of elite business actors who are often both global in their outlooks and orientation and *local* in their cultural predisposition and at the level of structural systems that are commonly *national* in their nature and organization (e.g. Hsu and Saxenian 2001; Olds 2001; Olds and Yeung 1999; Zhou and Tseng 2001). Revisiting the problem of culture, the literature on East Asian capitalism, particularly one that is associated with ethnic Chinese, is often replete with cultural essentialism. The role of *guanxi* or relationships in Chinese culture, for example, has been unproblematically treated as an exogenous and independent variable in explaining business behaviour of ethnic Chinese actors. This conflation of national culture emanating from mainland China and the everyday economic practice of ethnic Chinese in East and Southeast Asia has led to many serious misconceptions of the so-called '*guanxi* capitalism' (e.g. Redding 1990; cf. Hsu and Saxenian 2001). Ethnic Chinese actors in Asia have been seen as inward-looking and engaging in highly personalized transactions that undermine the market mechanism and open competition.

When we focus our analytical perspective on elite Chinese business actors, however, we can identify a whole array of their everyday economic practice ranging from corruptive activities of personalism to highly professional conduct of contracts and requisite legal processes. In other words, ethnic Chinese business elites in Asia are both local and traditional in their cultural norms and globalizing in their approach to business. In performing their capitalist organizations simultaneously at both global and local scales, these actors contribute to a dynamic process of Chinese capitalism morphing into a form of what Yeung (2004) calls hybrid capitalism. In this hybrid capitalism, there are both elements of culturally specific imprints (e.g. employment relations) and globalizing norms (e.g. international standards of corporate governance). Informed by Peter Dicken's (2003) work on major transformations in the global economy and Nigel Thrift's (2000) micro-examination of how culture is performed in capitalist firms, this concept of hybrid capitalism can capture adequately the complex duality of geographical flows (e.g. FDI) and economic landscapes in contemporary Asian economies.

Back to the future: key challenges for research and policy

What then does this review of economic geographers' work on Asian capitalism and their leading business actors (TNCs) mean for future economic-geographical research? To begin, we still know far too little about the dynamic transformations of regional economies outside North America and Western Europe. Most economic-geographical perspectives developed since the quantitative revolution in the 1960s have been situated in advanced industrialized economies, when at the same time other social science disciplines have been producing theories based on empirically-grounded research conducted in the developing world (Yeung and Lin 2003). This continual 'missing the boat' is perhaps one of economic geography's greatest contradictions. As an academic discipline that should be much more attuned to geographical differences, differentiation, and heterogeneity, economic geography has failed to deliver its verdict on a wide range of critically important research issues (e.g. the rise of China and India as economic superpowers). In the context of this short chapter, I can only outline five of them in order to drive urgent future research and policy agendas.

First, there is an urgent need to develop new theories that emanate from grounded research in economies outside North America and Western Europe. As a mix blend of hybrid capitalisms, the nature and dynamics of capitalist transformation occurring in Latin America, Eastern Europe, Africa, and Asia need to be much better theorized. Economic geographers should continue to be interested in firm-level analysis and pay more analytical attention to how business firms – indigenous and foreign – in these developing economies serve as key capitalist agents in bringing about technological change, economic spin-offs, and employment opportunities. The theoretical challenge to economic geographers is not so much about applying our existing analytical frameworks to these 'new' empirical problems, but rather about how we might develop genuinely grounded theories that in due course can help us 'theorize back' to economic-geographical problems found in advanced industrialized economies.

Second, there is now a much greater demand for comparative research that draws upon rich empirical insights into different forms of capitalist imperatives. In appreciating the manifold complexity of globalization tendencies, economic geographers can play an active role in the forefront of globalization research by examining how these global forces are impinging on different geographical realities at the same time (cf. Dicken 2004). This deep-seated concern for geographical differences and differentiation can only be accomplished in research terms through sustained comparative analysis. Whether we are concerned with industrial location, resource extraction, or business headquarters, we can always build in comparative analysis of how the same economic phenomenon works out differently in different geographical settings. In doing so, we can also develop better-grounded theories that account for these geographical differences.

Third, this chapter has clearly shown that economic geography might not seem to have a lot to offer on the globalization of Asian capitalism, a research topic that takes up some significant research efforts among political economists, sociologists and development scholars. While we may not feel comfortable to build alliances with conventional neoclassical economics (Amin and Thrift 2000), we should be prepared to establish more 'joint ventures' with such friendly disciplines as economic sociology, international political economy, comparative management, and so on that are analytically concerned with integral relations between economy and society. In building alliances with these interested disciplines, we should take care in maintaining the intellectual integrity of economic-geographical research. Too often an economic-geographical study may be accused for being sociological, political, or economic. In participating in multi-disciplinary initiatives, economic geographers must bring to the research table some useful analytical tools that are uniquely geographical, e.g. space, place, and scale. Without losing sight of our disciplinary identity, we do have something insightful to say about economic processes and institutions in the global economy.

Fourth and on the policy front, we can offer some useful suggestions for policy formulation in the context of reshaping Asian capitalism. Precisely because we focus on dynamics of change and adjustment, economic geography has much to inform the ongoing process of economic reform. Our appreciation of the complex interconnections of economic processes across different spatial scales allows us to offer policy suggestions that focus not just on national problems, but also on how economic issues are deeply spatial in their manifestation. While Asian business systems might be changing in the context of contemporary economic globalization, we can offer suggestions on economic policies that help to retain some culturally specific practices. Arguing against a wholesale adoption of 'global' standards of economic governance, we can make policy suggestions on economic reform that are much more attuned to local specificity and differentiation. One good example is the imposition of standard international accounting practice on Asian firms irrespective of their nature and organization. While greater transparency is generally good for global investors, it is important to understand the competitive dynamics of certain industries in Asia that might have a strong strategic outlook. This practice for greater transparency should be seen as an ideal state to be achieved gradually rather than an immediate task about sorting out the messy reality of Asian business.

Lastly, there are many useful policy implications for the future of globalizing Asian firms that economic geographers might offer. In particular, the unique trajectory to globalization charted by these Asian firms shows that there are indeed many ways to globalize (see Mathews 2002). There is no single market that automatically balances and arbitrages the demand and supply of globalization opportunities. In enhancing their firm-specific competitive strengths, many Asian firms do not necessarily need to rely on market-based mechanisms. Instead, the road to global competition and success is highly uneven and sometimes utterly unfair. This calls for selective and strategic intervention in the globalization trajectories of Asian firms by other capitalist institutions such as the state and non-state actors. This policy implication may not make sense in the context of the 'Washington consensus'. But in this world of neoliberal globalization, we can be sure that the condition of perfect market competition will never be satisfied and thus each firm and each economy needs to find its own way to economic prosperity and development. Ultimately, this process of economic development is necessarily different and uneven geographically.

References

Agmon, T. and C. P. Kindleberger (eds) (1977) *Multinationals from Small Countries.* Cambridge, MA: MIT Press.

Amin, A. and N. Thrift (2000) 'What kind of economic theory for what kind of economic geography?', *Antipode*, 32(1): 4–9.

Berger, P. L. and H. M. Hsiao (eds) (1988) *In Search of an East Asian Development Model.* New Brunswick, NJ: Transaction Books.

Cho, M. (1997) 'Large-small firm networks: a foundation of the new globalizing economy in South Korea', *Environment and Planning A*, 29:1091–108.

Coe, N., M. Hess, H. W. C. Yeung, P. Dicken and J. Henderson (2004) 'Globalizing regional development: a global production networks perspective', *Transactions of the Institute of British Geographers*, New Series, 29(4): 468–84.

Dicken, P. (1976) 'The multiplant business enterprise and geographical space: some issues in the study of external control and regional development', *Regional Studies*, 10: 401–12.

Dicken, P. (2003) *Global Shift: Reshaping the Global Economic Map in the 21st Century*, 4th edn. London: Sage.

Dicken, P. (2004) 'Geographers and "globalization": (yet) another missed boat?', *Transactions of the Institute of British Geographers*, 29(1): 5–26.

Dicken, P. and N. Thrift (1992) 'The organization of production and the production of organization: why business enterprises matter in the study of geographical industrialization', *Transactions, Institute of British Geographer*, New Series, 17: 279–91.

Dicken, P., P. Kelly, K. Olds and H. W. C. Yeung (2001) 'Chains and networks, territories and scales: towards an analytical framework for the global economy', *Global Networks*, 1(2): 89–112.

Grabher, G. (ed.) (1993) *The Embedded Firm: The Socio-Economics of Industrial Networks.* London: Routledge.

Grabher, G. (2006) 'Trading routes, bypasses, and risky intersections: mapping the travels of "networks" between economic sociology and economic geography', *Progress in Human Geography*, 30(2): 163–89.

Granovetter, M. (1985) 'Economic action, and social structure: the problem of embeddedness', *American Journal of Sociology*, 91(3): 481–510.

Hamilton, F. E. I (ed.) (1974) *Spatial Perspectives on Industrial Organization and Decision-Making*. London: John Wiley.

Heenan, D. A. and W. J. Keegan (1979) 'The rise of third world multinationals', *Harvard Business Review*, January–February: 101–9.

Henderson, J., P. Dicken, M. Hess, N. Coe and H. W. C. Yeung (2002) 'Global production networks and the analysis of economic development', *Review of International Political Economy*, 9(3): 436–64.

Hsing, Y. (1998) *Making Capitalism in China: The Taiwan Connection*. New York: Oxford University Press.

Hsu, J-Y. and A. Saxenian (2000) 'The limits of guanxi capitalism: transnational collaboration between Taiwan and the USA', *Environment and Planning A*, 32(11): 1991–2005.

La Porter, R. F. Lopez-de-Silanes and A. Shleifer (1999) 'Corporate ownership around the world', *Journal of Finance*, 54(2): 471–517.

Lall, S. (1983) *The New Multinationals: The Spread of Third World Enterprises*. Chichester: Wiley.

Leung, C. (1993) 'Personal contacts, subcontracting linkages, and development in the Hong Kong-Zhujiang Delta region', *Annals of the Association of American Geographers*, 83(2): 272–302.

Mathews, J. A. (2002) *Dragon Multinational: A New Model for Global Growth*. Oxford: Oxford University Press.

Mitchell, K. (1995) 'Flexible circulation in the Pacific Rim: capitalism in cultural context', *Economic Geography*, 71(4): 364–82.

Ohmae, K. (1990) *The Borderless World: Power and Strategy in the Interlinked Economy*. London: Collins.

Olds, K. (2001) *Globalization and Urban Change: Capital, Culture and Pacific Rim Mega Projects*. Oxford: Oxford University Press.

Olds, K. and H. W. C. Yeung (1999) '(Re)shaping "Chinese" business networks in a globalising era', *Environment and Planning D: Society and Space*, 17(5): 535–55.

Park, S. O. (1996) 'Networks and embeddedness in the dynamics of new industrial districts', *Progress in Human Geography*, 20(4): 476–93.

Qiu, Y. (2005) 'Personal networks, institutional involvement, and foreign direct investment flows into China's interior', *Economic Geography*, 81(3): 261–81.

Redding, S. G. (1990) *The Spirit of Chinese Capitalism*. Berlin: De Gruyter.

Thrift, N. (1996) *Spatial Formations*. London: Sage.

Thrift, N. (2000) 'Performing cultures in the new economy', *Annals of the Association of American Geographers*, 90(4): 674–92.

Vernon, R. (1971) *Sovereignty at Bay: The Multinational Spread of US Enterprises*. New York: Basic Books.

Wells, L. T. Jr (1983) *Third World Multinationals: The Rise of Foreign Investment From Developing Countries*. Cambridge, MA: MIT Press.

Whitley, R. (1992) *Business Systems in East Asia: Firms, Markets and Societies*. London: Sage.

Whitley, R. (1999) *Divergent Capitalisms: The Social Structuring and Change of Business Systems*. New York: Oxford University Press.

World Bank (1993) *The East Asian Miracle*. Oxford: Oxford University Press.

World Investment Report (2004) World Investment Report, available at: http://www.unctad.org/wir, accessed 15 June 2005.

Yang, Y. R. and C. J. Hsia (2006) 'Spatial clustering and organizational dynamics of trans-border production networks: a case study of Taiwanese IT companies in the Greater Suzhou Area, China', *Environment and Planning A* (forthcoming).

Yeung, H. W. C. (1994) 'Critical reviews of geographical perspectives on business organisations and the organisation of production: towards a network approach', *Progress in Human Geography*, 18(4): 460–90.

Yeung, H. W. C. (1997) 'Business networks and transnational corporations: a study of Hong Kong firms in the ASEAN region', *Economic Geography*, 73(1): 1–25.

Yeung, H. W. C. (1998) *Transnational Corporations and Business Networks: Hong Kong firms in the ASEAN region*. London: Routledge.

Yeung, H. W. C. (ed.) (1999) *The Globalisation of Business Firms from Emerging Economies*, two volumes, Cheltenham: Edward Elgar.

Yeung, H. W. C. (2000) 'The dynamics of Asian business systems in a globalising era', *Review of International Political Economy*, 7(3): 399–433.

Yeung, H. W. C. (2002) *Entrepreneurship and the Internationalisation of Asian Firms: An Institutional Perspective*, Cheltenham: Edward Elgar.

Yeung, H. W. C. (2003) 'Practicing new economic geographies: a methodological examination', *Annals of the Association of American Geographers*, 93(2): 442–62.

Yeung, H. W. C. (2004) *Chinese Capitalism in a Global Era: Towards Hybrid Capitalism*. London: Routledge.

Yeung, H. W. C. (2005) 'Rethinking relational economic geography', *Transactions of the Institute of British Geographers*, New Series, 30(1): 37–51.

Yeung, H. W. C. and G. C. S. Lin (2003) 'Theorizing economic geographies of Asia', *Economic Geography*, 79(2): 107–8.

Zhou, Y. and Y.-F. Tseng, (2001) 'Regrounding the "ungrounded empires": localization as the geographical catalyst for transnationalism', *Global Networks*, 1(2): 131–54.

Section III

Regional competitive advantage

Industrial change, human capital and public policy

13 Economic geography and the new discourse of regional competitiveness

Ron Martin

The new discourse of competitiveness

Although it may have had some earlier predecessors (see Reinert 1995), the term 'competitiveness' is really only a recent one. It entered general economic parlance in the mid 1980s, mainly through the writings of business school gurus, especially Michael Porter. But since then it has become a prominent, even hegemonic, discourse amongst policymakers the world over. Economists and experts everywhere have elevated 'competitiveness' to the status of a 'natural law' of the modern capitalist economy, and assessing a country's competitiveness and devising policies to enhance it have rapidly become officially institutionalised tasks.

What explains this new concern with competitiveness? There is little doubt that the popularity of the notion in policy circles is inextricably linked to the ascendancy and diffusion of pro-globalisation, pro-market neoliberal political ideologies among the advanced nations and many of their leading economic advisors. Under this credo, globalisation is not only an ineluctable process, it brings with it expanding trade and increasingly intense competition between firms and between nations (the 'threat' from India and China being increasingly invoked in this context), necessitating the pursuit of efficiency, flexibility and technological innovation in order to compete and survive in the global marketplace:

> A new era of competition has emerged in the last twenty years, especially in connection with the globalization of economic processes. Competition no longer describes a mode of functioning of a particular market configuration (a competitive market) as distinct from oligopolistic and monopolistic markets. To be competitive has ceased to be a means to an end; competitiveness has acquired the status of a universal credo, an ideology.
>
> (Group of Lisbon 1995: xii)

This new focus on competitiveness is by no means the sole preserve of neoliberal apologists, however; the belief that economic life in today's globalised and technologically-driven world is distinctly more 'competitive' has in fact gained widespread acceptance, even in left-of-centre political circles. The difference is that in the latter, 'competitiveness' (like globalisation) is often seen in a negative light, as an ultimately self-defeating imperative, whereas for the neoliberal it is a positive, indeed necessary feature of the free-market order.

An intriguing feature of this new discourse of competitiveness is that whilst initially a national-level concern, it has also stimulated considerable interest in regions and cities. One expression of this is a new policy emphasis on the 'regional foundations' of national competitiveness. In the United Kingdom for example, the Blair governments have repeatedly stressed the need to raise the competitiveness of the country's regions and cities in order to improve the nation's economic growth and productivity. Similarly, the European Commission sees the improvement of regional competitiveness across the Union as vital if it is to secure the goals set down in the Lisbon Agenda (of making the European Union the most dynamic knowledge-based economy by 2010):

> If the EU is to realise its economic potential, then all regions wherever they are located . . . need to be involved in the growth effort . . . Strengthening regional competitiveness throughout the Union and helping people fulfil their capabilities will boost the growth potential of the EU economy as a whole to the common benefit of all.
>
> (European Commission 2004: vii–viii)

Likewise, in the United States, research bodies such as the Washington-based Progressive Policy Institute and Harvard's Institute for Strategy and Competitiveness, have highlighted the importance of high-performing regions and cities for the competitiveness of the national economy. This new-found focus on regions and cities reflects a belief, again linked especially with neoliberal thinking, that the pursuit of 'competitiveness' requires close attention to the microeconomics of supply, and to the need to remove supply-side rigidities, barriers and related weaknesses in the economy. And this in turn, has promoted greater interest in regions and cities, where, it is believed, many supply-side problems reside and where policies aimed at their removal are best delivered and implemented.

At the same time, many regional and city authorities have themselves become increasingly concerned about the relative 'competitive standing' of their local economy compared to that of other regions and cities, and with devising strategies to move their area up the 'competitiveness league table'. Regional 'benchmarking', constructing rankings of regions and cities by this or that 'competitiveness index', has become a widespread practice. As globalisation has advanced, and nation-states have redrawn and withdrawn their spheres of economic intervention and regulation – or even lost some of their economic sovereignty to the onward march of globalising forces – so regional and city authorities see their local areas as both more exposed to the global economy and with greater autonomy to carve their own future within it. Comparing themselves with other 'competitor' regions and cities elsewhere has thus become one way of assessing their performance, their strengths and weaknesses.

All this resonates closely with the claim by many geographers (and others besides) that we are witnessing a (re)surgence of regions and cities as the loci of wealth production and economic governance in the world economy (see, for example, Best 2001; Ohmae 1995; Scott 1998, 2001; Storper 1997). How we conceptualise the regional and urban competitiveness is thus highly relevant to this alleged reassertion of regions, and economic geographers should, in principle, be well placed to provide some valuable insight. For the notion of 'place-' or 'territorial-competitiveness' would seem to be closely linked to what, traditionally, has been a central issue for economic geographers: namely, the pervasive phenomenon of geographically uneven development.

Yet, the idea of regional competitiveness is a contentious one, a notion around which there is no general consensus. Indeed, as Bristow (2005) puts it:

> Regional competitiveness lacks a clear, unequivocal and agreed meaning within the academic literature. It is perhaps not surprising therefore that the policy discourse around regional competitiveness is somewhat confused.
> (p. 289)

In fact, at the heart of this confusion are several questions. What, precisely, is meant by the term 'regional competitiveness'? In what sense do regions and cities compete? Are regions and cities meaningful economic units to which the notion of competitiveness can be meaningfully applied? Why should regions and cities differ in competitiveness? What are the policy implications of regional and urban differences in competitiveness? Policy concerns with urban and regional competitiveness have run ahead of answers to these and related questions. A substantial research effort would thus seem to be called for to redress this imbalance and provide a firmer base for policy debate.

Competitiveness: a contentious concept

One source of confusion is that even in economics, the idea of 'competitiveness' has attracted considerable debate. For the individual firm it is often taken to mean the ability to create, retain or expand market share for some product or service, on the basis of price, quality, design, delivery, or some other advantage. Firms that progressively lose market share and face declining profitability are deemed to be 'uncompetitive', and may ultimately go out of business. But what does the term mean for economic aggregates above the level of the firm? At the national scale, definitions have proliferated (see Cellini and Soci 2002), prompting an early critical salvo by Reich (1990) to the effect that: 'National competitiveness is one of those rare terms of public discourse to have gone directly from obscurity to meaninglessness without any intervening period of coherence'.

This is not entirely true, however, since most definitions of national competitiveness refer in some way or another to a nation's economic 'performance', be this Gross Domestic Product (GDP) per head, productivity, or trade balance. Frequently, reference is made to a 'nation's ability to produce goods and services that meet the test of international markets, while at the same time maintaining high and sustainable levels of income and employment'. Yet for writers like Paul Krugman (1996a, 1996b) this appeal to trade performance is itself problematic, since it can all too easily conjure up a neo-mercantilist image of nations competing one against another, in a zero-sum fashion, over shares of particular product or service markets. According to Krugman, the notion of competitiveness is an attribute of firms but not of cities, regions or nations. Others disagree. Michael Porter in his seminal studies of 'competitive advantage' deplores the lack of attention to competitiveness in economic analysis (Porter 1990, 1998). He goes on to argue that the national environment affects the competitive position of firms, and that understanding that environment would yield some fundamental insights into how competitive advantage at the firm level is created and sustained (1990: xii).

But if defining the concept of 'competitiveness' at the national level is contentious, it is doubly so at the regional or local scale. For one thing, some geographers would argue that confusion surrounding the notion of 'regional competitiveness' also arises because the concept of the 'region' itself is equally problematic. It may be that regions have become increasingly salient loci in the global economy, but defining and conceptualising regions, it is contended, has simultaneously become increasingly more complex – in part because of the very globalisation that is promoting the new discourse on competitiveness. The problem is that regions are typically not pre-given, fixed, internally-coherent economic units, but highly fuzzy, open and internally discontinuous entities, the various spatial and economic components of which are differently linked into different aspects of the both the national and global economy. There is no pre-existing, singular 'essential' geographical economic space called the 'region': rather there are different regional representations of economic space depending on the specific issue under enquiry and the perspective adopted (Allen et al. 1998: 34). In addition, there is the issue of agency. Regions are not decision-making entities in the same way that firms are, but instead consist of 'bundles' of firms, organisations, social groups and institutions, all with their own imperatives, dynamics and networks of interactions. And regional authorities typically have little or no direct control or influence over the firms within their areas. Hence, many geographers would have reservations about the idea of 'regional' competitiveness.

However, just as in a Coasian view of the world, where it is the organisation of productive assets in a firm that gives rise to the analysis of the firm as a unit of production, so nations, regions and cities too can be seen as collections of assets, variously organised, so that it is reasonable to think in terms of the competitiveness of that bundle of assets, even if Krugman is right in advocating caution about making analogies between the firm and the nation or region. Furthermore, although most regional units used for policy and analytical purposes are based on political or administrative boundaries that need bear little correspondence to economic relationships, there are certain features about such 'official' regions that do give them some measure of meaning as economic entities. Thus regional authorities often have tax-raising powers and responsibilities for spending on public services, utilities and infrastructure, all of which impact on local firms. Also, as noted above, regional authorities and bodies are becoming increasingly active in other areas of local economic governance, whether as the delivery agents of decentralised national government policies, or as active policy agents in their own right and capacity. It may be that regions are difficult to define as 'essential' economic units, but the fact is that a process of 'regional institutionalisation' of policy intervention and responsibility appears to be underway that is endowing politically and administratively defined regions with some degree of functional economic meaning. It is as part of this institutionalisation process that regional authorities and bodies are busy devising policies to improve and upgrade the competitiveness and productivity of the businesses, workers and organisations in their jurisdictions. If only because of this rise of the region as an arena of economic governance and intervention, and the increasing trend for policymakers to think of regions as the sites of competitive advantage, it is important to appraise the different senses in which the term 'regional competitiveness' is used.

There are in fact two interrelated questions that research needs to address: does thinking in terms of competitiveness throw light on how we define and analyse regional economies? And does a regional (geographical) perspective help us to understand competitiveness? Both questions are worthy of serious attention by economic geographers, and both have direct policy implications.

Thinking about regional competitiveness

Empirical observation amply testifies to the fact that some cities and some regions (however defined) do better – in terms of average prosperity, employment, standard of living, growth or some other measure of 'performance' – than others. Geographers have long highlighted spatial disparities and uneven development of this sort. Geographers have not traditionally thought of such disparities in performance explicitly in terms of competitiveness, although notions of 'place competition' have woven their way through the economic geography literature. For example, much of traditional location theory, in economic geography and in regional science, was concerned with deriving the spatial structure of the economy as the outcome of a particular model of competition (such as perfect competition), under specific assumptions as to the production function of firms (especially the assumption of diminishing returns to inputs), the geographical distribution of resources and consumers, transport costs, and the movement of labour and capital between places. Given that the dual focus of much of this work was on the 'relative attractiveness' of locations to firms and workers and on inter-firm competition across space (Sheppard 2000), the implication was that locations and places do 'compete' in some sense, for example for capital, labour and markets. Nevertheless, overall, the main aim of this work was on explaining the location of industry and deriving equilibrium economic landscapes of activities, markets and prices, not with unravelling the nature of regional or place competitiveness as such.

Similarly, Marxian economic geography also reverberates with implicit notions of 'competition between places'. One of the key arguments of this approach was that in response to changes in technology, costs, and market conditions, capital constantly shifts from region to region in order to exploit geographical variations in the opportunities for profitability. In seeking the most profitable locations in this way, capital in effect 'plays off' different regions according to their relative advantages for accumulation, so that development in certain regions tends to be at the expense of that in others. Again, in a sense, regions are seen as competing one against another. Further, this process is viewed as being relentless, denying the creation of an equilibrium economic landscape, and continually reshaping the relative advantage of different places as far as capital is concerned. As in the case of location theory, however, the notion of regional competitiveness is not itself the focus of analysis.

It is only in the last few years that the subject of 'place-' or 'territorial-competitiveness' has begun to attract serious attention in its own right (see, for example Begg 2002; Boschma 2004; Bristow 2005; Camagni 2003; Kitson et al. 2006; Krugman 2003; Malecki 2004; Porter 2001; Storper 1997; *Urban Studies* 1999). But, somewhat ironically, it has not been geographers but economists – especially those that have 'gone geographical', notably Michael Porter and Paul Krugman – who have led the new discourse of regional and urban competitiveness and brought the idea to the attention of policymakers.

According to Porter (2001) a 'new economics of competition' is emerging that is associated with six transitions: from macroeconomic policies to microeconomic policies that recognise that the 'drivers' of prosperity are based at the sub-national level; from a concern with current productivity to emphasising innovation, as the basis of sustained productivity growth; from the economy as a whole as the unit of analysis to a focus on 'clusters' (groups of interlinked specialised activities, often geographically localised); from internal to external sources of company success, recognising that the location of a company can affect the capabilities it can draw upon; from separate to integrated economic and social policy; and from national to regional and local levels as the locus of analysis and policy intervention. Indeed, in Porter's view, economic geography assumes a pivotal role in understanding this 'new competition':

> The more that one thinks in terms of microeconomics, innovation, clusters and integrating economic and social policy, the more the city-region emerges as an important unit. Issues or policies that span nations or are common to many nations will be increasingly neutralised, and no longer sources of competitive advantage. However, it is not a matter of one unit of geography supplanting another . . . The task is to integrate the city-region with other economic units, and to adopt a more textured view of the sources of prosperity and economic policy that encompasses multiple levels of geography.
> (ibid.: 141)

As for Krugman, in what is a major departure from his previous dismissal of 'competitiveness talk', he now argues that the notion may after all have particular relevance at the regional level

> Success for a regional economy, then, would mean providing sufficiently attractive wages and/or employment prospects and return on capital to draw in labour and capital from other regions. It makes sense, then, to talk about 'competitiveness' for regions in a way one wouldn't talk about it for larger units.
>
> (Krugman 2003: 19)

Underpinning these discussions of regional and city competitiveness by economists such as Porter and Krugman is a rethinking of trade and a rediscovery of increasing returns.

Standard (mainstream) economic theory would suggest that the capacity of a region to compete is shaped by an interplay between the attributes of region (or cities) as locations and the strengths and weaknesses of the firms and other economic agents active in them. If markets worked perfectly, it might be expected that inter-regional cost differentials would adjust to give rise to a pattern of regional trade in which comparative advantage (differences in factor endowments) determined relative specialisation and trade amongst regions (and cities). However, the persistence of regional differences in key economic indicators (such as incomes and employment) suggests that there are systematic differences in the relative attractiveness of different regions and cities. We know for example that movements of capital and labour are not such as to eliminate differences in costs and returns between regions. Thus, as Krugman argues, the ability of a region or city to attract capital and labour is both a measure of its competitiveness as a location and a source of cumulative competitive advantage for that region or city. This point is also emphasised by the geographer Michael Storper, who defines regional competitiveness as:

> the ability of a regional economy to attract and maintain firms with stable or rising market shares in an activity while maintaining or increasing standards of living for those who participate in it.
>
> (Storper 1997)

A not dissimilar focus on the relative attractiveness of places is also to be found in Richard Florida's work, which argues that it is not so much the ability of places to attract firms, but rather their ability to attract and retain 'creative people' that matters for regional growth, so that places compete as locations for such people, and their success in attracting them in turn shapes the relative competitive performance of those places. Florida's theory is that:

> Regional economic growth is driven by the location choices of creative people – holders of creative capital – who prefer places that are diverse, tolerant and open to new ideas . . . It identifies a type of human capital, creative people, as being key to economic growth, and . . . it identifies the underlying factors that shape the location decisions of these people, rather of merely saying that regions are blessed with certain endowments of them.
>
> (Florida 2000: 223)

In addition, it now accepted that Ricardian comparative advantage (relative factor endowments) theory is not the only basis for trade between nations, and that increasing returns within industries also play a key role. Economists have also recognised that a nation's globally competitive industries, those in which it has a trading advantage, often tend to be geographically concentrated in particular regions and localities. This has led to the acknowledgement that many of the increasing returns in contemporary economic development are regional or local in origin, in the form of external economies associated with the geographical agglomeration of industry and with neo-Marshallian districts or clusters of economic specialisation, and that to understand such issues as trade, competitiveness, innovation and productivity, we need to examine these local externalities.

Economic geographers have of course drawn equally heavily on external economies ideas in recent years. Whether the terminology used is that of industrial districts, clusters, or local production systems, the argument is that the close geographical proximity or spatial agglomeration of similar and related firms (often, but not always, in some complex inter-firm division of labour) enables firms to benefit from a range of locally-emergent and locally-embedded externalities, such as access to specialised labour and specialised suppliers, spillovers of technology and knowledge, specialised institutions, and networks of trust and shared business cultures and practices (or what geographers have come to call 'untraded interdependencies'). These local externalities, in one form or another, are now widely argued by geographers to be an important basis of regional economic success. Not only do such externalities influence the nature and extent of 'collective learning' and innovation amongst the firms (and institutions) making up an industrial district or cluster (Maskell et al. 1998), they also shape the attractiveness of the locality to other similar firms, workers and institutions. In short, favourable local externalities raise the ability of local firms to compete, whilst also attracting firms, workers, and ideas from elsewhere.

What these economic and geographical literatures on localised externalities point to is the importance of locally emergent and embedded effects in the shaping of regional competitive advantage. Indeed, Krugman (2003) identifies two primary sources of regional competitiveness: 'regional external economies', and 'regional fundamentals':

> Given that modest differences in total factor productivity can have large growth consequences at the regional level, what accounts for such differences? A broad division would be between 'fundamentals' – differences rooted in a region's characteristics – and 'external economies' that are themselves a consequence of a region's pattern of economic development.
>
> (pp. 23–4)

Regional external economies are the (neo-Marshallian and related) localisation economies referred to above. These emerge as a consequence of the development of geographical concentrations of specialised or interrelated activities. They form part of the local 'environment' (to use Porter's terminology) on which firms draw in creating and sustaining their competitive advantage. Krugman uses the term regional fundamentals to refer to key regionally-embedded assets such as a well educated workforce, the result of a strong tradition of good schooling and higher education, a local culture of entrepreneurship, modern infrastructure (physical, social and cultural), and a sustained form of purposive public policy. These also form part of the local environment that influences the competitive performance of a region's firms. He tends to treat these two sources of regional competitiveness as distinct. In reality, of course, they are likely to be interrelated and mutually reinforcing: the particular economic clusters and specialisations that develop in a region are – to some extent – influenced by a region's fundamentals, and the latter will in turn be (re)shaped by those specialisations and the externalities that emerge around them.

What Krugman calls regional fundamentals are similar to what Camagni (2003), in his discussion of territorial competitiveness, refers to as regional 'absolute advantages', the region-specific sets of socio-institutional and cultural norms, networks and structures (including social capital) that impact on economic outcomes (see Granovetter 2005). Such characteristics tend to be relatively immobile between regions, and hence are sources of both relative and absolute advantage. Economic geographers now put considerable emphasis on the role of local 'institutional thickness' and social embeddedness in local economic development, though demonstrating their importance has proved far from straightforward. For example, the empirical evidence for the impact of social capital on economic performance, nationally or regionally, remains somewhat ambiguous (see, for example, Casey 2004). Nevertheless, the notion of regional fundamentals does at least highlight the way in which a region's 'assets' extend well beyond its stock of firms, and their traded and untraded interdependencies, and its labour force, to include a host of social, institutional and cultural features that both shape regional economic development and serve to some extent to differentiate one regional economic space from another.

Such considerations suggest, then, that we can think of regional competitiveness in at least two, interrelated ways: in terms of a region's relative attractiveness as a location for mobile labour, capital and knowledge; and in terms of the relative range, quality and nature of a region's externalities and fundamentals that together constitute a locally-specific environment of resources, capabilities or assets on which local firms and workers can draw, directly or indirectly. This latter idea seems to be behind the European Commission's argument that the idea of regional competitiveness:

> Should capture the notion that, despite the fact that there are strongly competitive and uncompetitive firms in every region, there are common features within a region which affect the competitiveness of all firms located there.
>
> (European Commission 1999: 5)

Of course, a region's 'resource environment' (or 'common features') can impact either favourably or unfavourably on its economic development; and an environment that was once highly favourable can at some later date become progressively less favourable, as once positive externalities and fundamentals cease to be sources of competitive advantage. Indeed, one of the most intriguing aspects of regional competitiveness is why some regions are able to maintain their relative competitive advantage over long periods of time, while others lose theirs and then find it difficult to recreate it. The basic point is that regional competitiveness is a dynamic process, subject to constant pressures and changes from both without and within. How a region's economy reacts and adjusts to such changes is really at the heart of the regional competitiveness issue.

Regional competitiveness as an evolutionary process

The 'dynamic adaptive capability' of regional economies is therefore of central importance. By this is meant the capacity of a region's firms, industries, and institutions to sense opportunities (market, technological, organizational), to nurture, adapt and regenerate their knowledge assets and competences, and to develop and enhance the organizational capabilities that translate that knowledge into effective actions. This general notion applies to individual firms, to whole industrial sectors, to social and public institutions, and to policy-making bodies alike. It reflects the capacity of firms to experiment with and shift to new product-specific capabilities; for industrial sectors it has to do with the success with which the firms in that sector are able to move into new markets, or upgrade existing ones; it has to do with the capacity of local entrepreneurs to identify and venture into new products and technologies; and it has to do with the capacity of institutions of all kinds to be receptive to change and new opportunities. In short, the greater the dynamic adaptive capability of a region's economy and socio-institutional base, the more likely it is to maintain or enhance its relative competitive performance over time.

In other words, regional competitiveness should be seen as an evolutionary process (Boschma 2004). Economic geographers have barely begun to explore the full scope of 'evolutionary economics' (of which there are several different variants, including neo-Schumpeterian, institutionalist, game-theoretic, and complexity-theory based, for example), but it is clear that evolutionary economics contains several concepts, analogies and metaphors that bear directly on the definition and explication of regional competitiveness. Evolutionary theory forces us to think carefully about what economic competition means, the basic economic units that evolve – such as firms, routines, institutions – and what the mechanisms of regional structural, technological and institutional change are.

Essentially, economic evolution is about innovation and adaptation, and how these drive the direction and nature of structural change. Understanding the processes that determine patterns of innovation and adaptive structural change across regions should therefore throw valuable light on why regions differ in competitive advantage, and moreover, how and why patterns of regional competitive advantage shift and change over time. Economic geographers put particular emphasis on innovation as a source of regional performance. But what determines regional differences in innovation? Evolutionary theory stresses the importance of variety as a source of novelty. This accords with a Jacobsian view of economic change, whereby innovation is promoted by local economic diversity and heterogeneity, since this maximises both the scope for interaction and the variety of market opportunities for new ideas. Many successful large city-regions fit this model. In contrast, numerous economists and economic geographers have tended towards the Marshall-Arrow-Romer view that innovation is stimulated by local economic specialisation, where it is driven by intense rivalry between, and knowledge spillovers amongst, local firms in the same industry (or in closely related industries). This is essentially the assumption employed by Porter in his cluster model of regional competitive advantage, and by economic geographers in their studies of high-technology regions and districts. Certainly some successful regions and localities fit this model, although economic landscapes everywhere are littered with old specialised regions and localities that were once innovative leaders, but which have long since lost their prominence and are today's problem areas. So local economic specialisation is not of itself a guarantee of sustained competitive advantage.

This relates to another central idea of evolutionary economics that is highly pertinent to the question of dynamic regional competitiveness, namely that of path dependence. The concept of path dependence is intended to capture the process by which the evolution of the economy is always the contingent outcome between change and inertia. Economic choices and opportunities are always conditioned to some extent by dependence on past structural, institutional, social and technological developments. The economy is an irreversible historical process, in which at any point in time the state of the economy depends on the historical adjustment path taken to it. Technology and institutions are two of the primary 'carriers of history' that result in path dependence. And both are characterised by tendency for 'lock-in', that is for particular patterns of behaviour, technological organisation, economic specialisation, institutional arrangements and the like to become self-reproducing over time, despite other possible patterns, activities and arrangements. The neo-Marshallian and related local external economies referred to earlier tend to impart such lock-in, as do other forms of inter-relatedness amongst local firms, sunk costs, and institutionalised social routines and networks. Lock-in, in fact, is a pervasive feature of socio-economic life. In a regional context, the emergence of economic, technological, social and institutional structures can be heavily dependent on local context, but once established, the very interactive, situated and continuity-preserving nature of socio-economic activity is such that there are likely to be a tendency for the selected structures to get 'locked in'. Regional economies everywhere inherit the legacy of their past development. Geographers invoking the concept of 'lock-in' have invariably tended to ascribe negative or sub-optimal connotations to it, to view it as a barrier to change – the 'weakness of strong ties' argument. But this is too one-sided a reading: 'lock-in' can also be a positive feature, the source of increasing returns and competitive advantage. Indeed, this is how self-reinforcing development is typically initiated, and almost every regional economy – highly successful as well as less prosperous – displays attributes and examples of lock-in. What matters is why and under what circumstances lock-in turns from being a positive process into a negative one, and why this varies across regions, how some regions have proved better able to escape negative lock-in and to foster new paths of development and competitive advantage: in short, why some regional economies are more adaptive than others.

By viewing regional competitiveness as an evolutionary process, considerable scope is opened up for the application and extension of key ideas from evolutionary economics – such as adaptation and path dependence – within economic geography and regional theory more generally.

Regional competitiveness and policy research

Similarly, the study of regional competitiveness also opens up opportunities for a greater engagement by geographers with public policy research and debate, of the sort argued for by some commentators (such as Markusen 1999; Martin 2001). As noted above, regional competitiveness policy has tended to rush ahead of theoretical understanding and the evidence base. Economic geographers can make valuable contributions on both fronts. There is a pressing need to constructively interrogate the meaning and nature of 'regional competitiveness', both to provide a firmer base for understanding regional differences in economic success and for informing policy discourse. A geographical-theoretic perspective, for example, would not only highlight the importance that place makes to economic organisation and performance, and thus how local context matters even in an increasingly global world, but also – to pick up the argument made by Porter – how the processes influencing competitive advantage operate and interact at various spatial scales. It would also highlight the need to include intra-regional (or intra-urban) socio-spatial distributional issues into any definition and analysis of regional or city competitiveness.

Economic geographers are likewise well placed to engage directly with policy discourse, not only because competitiveness policy is itself increasingly regional and city-based, but because such policies are often predicated on an explicit comparative argument, involving direct comparisons between individual regions and cities. Geographical research can help reveal the scope for and limits to this 'benchmarking' and use of 'exemplar' places that seems now to be an essential part of competitiveness policy at national, regional and city levels. Certainly, if done properly, regional benchmarking can help identify a region's or city's competitive strengths and weaknesses, and hence form the basis of policy formulation and priorities. It can help mobilise and articulate the interests of the key actors and groups in the regional economy: the local business community, workers, and public and private institutions. And it can help a region's business, political and social communities forge a common sense of purpose in terms of ambitions for the future, and in presenting the region to the global market place, even in lobbying efforts to influence Government policies and the allocation of resources. Regional benchmarking can facilitate the development and ongoing review of a vision defining the region's role in a world economy characterised by a steadily increasing and ever-shifting division of labour.

But such benchmarking is fraught with dangers and limitations. What precisely does it mean to compare one city, one region, with another? While it is certainly instructive to examine and learn from successful regions, policymakers should be wary about treating them as exemplars that can be easily replicated or imitated in their own region. Policies rarely travel well: successful strategies developed in one region need not transplant easily into other regions (especially in other countries). Indeed, given that many of the sources of regional competitive advantage are locally based and embedded, policies necessarily have to respond to, and take account of, regionally-specific circumstances. Together with the problems in defining, measuring and explaining regional competitive advantage discussed in this chapter, it follows that there is unlikely to be any 'one size fits all' strategy for enhancing regional competitiveness. Different regions will face different problems, different types of competition, and require somewhat different policy mixes and emphases. Economists prefer universal tendencies and transferable policies: economic geographers have a comparative advantage in recognising and demonstrating the difference that place makes.

Whether we like it or not, whether we agree with it or not, competition is an integral feature of economic, political, social and cultural life. It is not simply a neoliberal invention. Economic geographers have an important role to play in elucidating the nature of and limits to the idea of 'regional competitiveness', as a way of thinking about the economic landscape, as an empirical process, and as a form of policy thinking.

References

Allen, J., A. Cochrane and D. Massey (1998) *Rethinking the Region*. London: Routledge.

Begg, I. (ed.) (2002) *Urban Competitiveness: Policies for Dynamic Cities*. Bristol: Policy Press.

Best, M. (2001) *The New Competitive Advantage*. Oxford: Oxford University Press.

Boschma, R. (2004) 'Competitiveness of regions from an evolutionary perspective', *Regional Studies*, 38: 1001–14.

Bristow, G. (2005) 'Everyone's a "winner": problematising the discourse of regional competitiveness', *Journal of Economic Geography*, 5: 285–304.

Camagni, R. (2003) 'On the concept of territorial competitiveness: sound or misleading?', *Urban Studies*, 39: 2395–411.

Casey, T. (2004) 'Social capital and regional economies in Britain', *Political Studies Quarterly*, 52: 96–117.

Cellini, R. and A. Soci (2002) 'Pop competitiveness', *Banca Nazionale del Lavoro, Quarterly Review*, 55(220): 71–101.

European Commission (1999) *Sixth Periodic Report on the Social and Economic Situation of Regions in the EU*. Brussels: European Commission.

European Commission (2004) *A New Partnership for Cohesion: Convergence, Competitiveness and Cooperation*. Brussels: European Commission.

Florida, R. (2002) *Rise of the Creative Class.* New York: Basic Books.
Granovetter, M. (2005) 'The impact of social structure on economic outcomes', *Journal of Economic Perspectives,* 19(1): 33–50.
Group of Lisbon (1995) *Limits to Competition.* Cambridge, Mass: MIT Press.
Kitson, M., R. L. Martin, and P. Tyler (eds) (2006) *Regional Competitive Advantage.* London: Routledge.
Krugman, P. (1996a) 'Making sense of the competitiveness debate', *Oxford Review of Economic Policy,* 12: 17–25.
Krugman, P. (1996b) *Pop Internationalism.* Cambridge, MA: MIT Press.
Krugman, P. (2003) *Growth on the Periphery: Second Wind for Industrial Regions?* Strathclyde: Fraser Allander Institute.
Malecki, E. (2004) 'Jockeying for position: what it means and why it matters to regional development policy when places compete', *Regional Studies,* 38: 1101–20.
Markusen, A. (1999) 'Fuzzy concepts, scanty evidence and policy distance: the case for rigour and policy relevance in critical regional studies', *Regional Studies,* 33: 869–84.
Martin, R. L. (2001) 'Geography and public policy: the case of the missing agenda', *Progress in Human Geography,* 25: 121–37.
Maskell, P., A. Malmberg, E. Vatne, H. Eskelinen and I. Hannibalsson (1998) *Competitiveness, Localised Learning and Regional Development.* London: Routledge.
Ohmae, K. (1995) *The End of the Nation State: The Rise of Regional Economies.* London: HarperCollins.
Porter, M. E. (1990) *The Competitive Advantage of Nations.* New York: The Free Press.
Porter, M. E. (1998) *On Competition.* Cambridge, Mass: Harvard University Press.
Porter, M. E. (2001) 'Regions and the new economics of competition', in A. J. Scott (ed.) *Global City Regions,* pp. 139–52. Oxford: Oxford University Press.
Reich, R. (1990) 'But now we're global', *The Times Literary Supplement,* 31 August.
Reinert, E. S. (1995) 'Competitiveness and its predecessors – a 500-year cross-national perspective', *Structural Change and Economic Dynamics,* 6: 25–47.
Scott, A. J. (1998) *Regions and the World Economy: The Coming Shape of Global Production, Competition and Political Order.* Oxford: Oxford University Press.
Scott, A. J. (2001) *Global City Regions.* Oxford: Oxford University Press.
Sheppard, E. (2000) 'Competition in space and between places', in E. Sheppard and T. Barnes (eds) *A Companion to Economic Geography,* pp. 169–86. Oxford: Blackwell.
Storper, M. (1997) *The Regional World.* New York: Guilford Press.
Various (1999) 'Competitive Cities', *Urban Studies (Special Issue),* 36(5/6).

14 Economic geography as (regional) contexts

Bjørn T. Asheim

Introduction: geography as context

Context is important for understanding. Geography, according to my PhD supervisor at Lund University, the famous Swedish geographer Torsten Hägerstrand, is about doing contextual analysis as opposed to compositional analysis, which is the task of other scientific disciplines (Hägerstrand 1974). This distinction corresponds to the one the German philosopher Immanuel Kant used when classifying sciences either as physically or logically defined. Geography and history understood as chorology and chronology respectively constitute the physically defined sciences, while other disciplines are logically defined based on their respective objects of study. Geography and history are synthetic (i.e. empirical based) sciences, while the logically defined are analytical. These distinctions are in my view fundamental in understanding the *raison d'être* of geography as well as its place and position in the division of labour with other disciplines.

Looking specifically at human geography and the whole history of ideas of the subject, the last 70–80 years can be interpreted as a struggle between a traditional position of geography as an idiographic, physically defined discipline (i.e. regional geography), others wanting to turn human geography into a nomothetic, analytical discipline, and later attempts trying to develop a theoretical informed, contextual approach transcending the idiographic-nomothetic dichotomy. The nomothetic position was primarily represented by 'spatial analysis' defining the object of study of geography as 'space' (i.e. 'spatial patterns' and 'spatial processes'), leaving 'history' to history and 'society' to the other social sciences and, thus, finding a place for human geography among the analytical social sciences (Schaefer 1953). As will be discussed later, this position was neither unproblematic nor sustainable in the long run for a social science, even if it had a hegemonic position until the demand for 'social relevance' started to be voiced loudly at the end of the rebellious 1960s.

Personal and educational background – and early years of research experience

Also personally and educationally a contextual perspective promotes understanding. Trained as a business economist with a broad background in business administration, economics, economic geography and economic history (MSc from the Norwegian School of Economics and Business Administration in Bergen) I already in my early years got a substantial (in contrast to only a formal) understanding of economics due to economic geography and history, even though the teaching in economics was mainstream.[1]

After graduating in 1971 I continued studying regional economics and geography before starting as a research assistant in a governmental research project called 'The Level of Living Study' in mid 1972. Being the only researcher with some background in economic geography I got the responsibility of handling the regional part of the study together with the senior researcher within this area. This was not an easy task, as very little had been done or written about regional inequalities in level of living. David Smith has just started doing some research in the United States before moving to South Africa, Richard Morrill had published a few small articles in *Antipode* as had Anne Buttimer, however, these initial attempts in the beginning of the 1970s did not provide a lot of support for a young researcher. *Antipode*, of which I was the first and for a long time the only subscriber in Norway, was my main source of inspiration.

My own (level of living) research taught me a couple of very important lessons. When presenting preliminary results from my study at a graduate seminar at the end of 1972 at the (common) geography department of the Norwegian School of Economics and the University of Bergen I was told by the professors that this was not geography but sociology, as it studied regional aspects or dimensions of social problems and processes, and not only spatial processes as the 'spatial analysis' tradition said. This in fact was the first time that the 'social relevance' discussion, which in an Anglo–American context was introduced at the end of the 1960s, was raised in Norway. In this way it can be argued that I actually brought the discussion of 'social relevance' to Norwegian geography. In later writings David Smith (1979) stated explicitly that the perspective of geography of welfare or social well-being made it necessary to have the social as the starting point, as welfare and well-being was fundamentally a social and not a spatial phenomenon. Space can never be the starting point for theoretical work within social sciences.

This, of course, concerns the key problem in geography of the space-society relationship or the adequate level of the theoretization of space. Geography as *chorology* traditionally implied an analytical distinction between space and society, defined as a non-spatial entity, which was studied by other social scientists (e.g. economists). In the 'spatial analysis' tradition, dominating economic geography until the beginning of the 1970s the explicit object of study was the spatial and the ambition was analytical, as was indicated by the title of some of the seminal contributions of this period, *Location Analysis*, by Peter Haggett (1965) and *Spatial Organization* by Abler et al. (1971). While clearly representing a scientific progress moving from descriptive and idiographic regional geography studies to theoretical and nomothetic spatial analysis, at the end of the 1960s – paradoxically around the time when David Harvey published his methodological bible on positivist spatial analysis (1969), *The Explanation in Geography* – this tradition had stiffened in empty, formal analyses, using tools developed by the 'quantitative revolution' in geography,[2] of the appearances of spatial phenomena as such independent of the social, economic and political importance of the events studied. This approach could neither survive the political radicalization of the student population after 1968 nor the critique of positivism in the social sciences (which also turned up at a later stage in human geography compared to other social sciences), and a strong demand for more 'social relevance' in the discipline was the result.

The demand for 'social relevance' influenced human geography in many ways, and resulted in the appearance of several new directions. In addition to the level of living or welfare geography studies, which clearly was a response to the previous lack of 'social relevance' focusing on real social questions,[3] a radical approach, which came to mean a Marxist based approach, to economic geography was the most prominent. I became associated with this approach in the early 1970s through contacts with young Danish geographers at the geography departments at Copenhagen University and the newly established (1972) Roskilde University Centre (RUC) just outside Copenhagen. For the rest of the 1970s I was the only Norwegian Marxist geographer. Danish human geography had been extremely traditional, and the young generation graduating around the time when the idea about 'social relevance' diffused, looked to an East German geographer, Schmidt-Renner, for inspiration. The outcome of these efforts was that Denmark became one of the strongholds of Marxist human geography in the 1970s outside the Anglo-American world, with radical milieus at all three geography departments (in addition to the two above mentioned also at Aarhus University). They formulated what was to be known as the 'territorial structure' geography. My contacts with this milieu were strengthened when moving to Lund University in 1976 to start on my PhD degree. In the autumn of 1978 I was employed as an external lecturer at RUC to teach the history of geographic thought to graduate students. In the spring of 1979 I became associate professor in human geography at Aarhus University approximately around the same time as I defended my PhD dissertation (May 1979) on *Regional inequalities in level of living*. My contacts with graduate students at Roskilde and Aarhus, who in most cases were Marxist oriented economic geographers, forced me to speed up my reading of Marx to be able to give competent supervision. The main focus of the students' work was the analysis of technological change in a capitalist mode of production. At this time the Marxist frame of reference (especially at RUC) had moved away from the rather orthodox historical materialist interpretation of the territorial structure geography to what is known as 'west-European left-Marxism'.

Characteristic for this tradition is a history of ideas approach to the background and development of Marx' thought. Of special importance is the highlighting of the importance of the dialectical, philosophical thinking, derived from Hegel, in Marx' political economy work. Moreover, in contrast to more traditional interpretations this approach differentiates between (Asheim and Haraldsen, 1991): (a) three different phases in Marx' writing (the young Marx up; the period with historical-materialist works (1845–57); and the period to his death in 1883 in which the central works on the critique of the political economy (*Grundrisse* and *Capital*) were written), implying that the work of the different phases cannot be regarded as identical theoretical projects; (b) logic and history referring to the different levels of abstraction in Marxist theory (i.e. theories about the logic and laws of motion of capital found in *Capital* vs. studies of concrete social formations). In the studies of technological change this differentiation underlines the interrelation between the (exchange) value dimension (economy) and the material (use value) dimension (technology), implying that the capitalist production process is a valorization as well as a labour process, where the valorization process subsumes that labour process (Asheim 1985). This makes simplistic explanations of, for example, locational changes deduced from changes in the valorization process impossible, and establishes studies of concrete social formations as a specific level of analyses in a Marxist theoretical approach. Also this alternative approach emphasizes that Marx gave up the paradigm of necessity in his political-economical works. This means that the logic of capital must be interpreted as tendencies (i.e. necessary, internal relations of the capitalist mode of production), which implies that it is not a question of things being predetermined, but only determined by the tendencies (structures) whose realization are dependent on contingently related conditions. In many ways this approach provides answers to most of the criticism Marxism was exposed to, for example, in the debate in *Environment and Planning D: Society and Space* in 1987, which represented the ending of the hegemonic position of Marxist economic geography.

This non-deductive and non-reductionist approach represented some serious methodological challenges, which could not easily be answered by looking for methodological guidelines in Marx' own writings. Beyond referring to the 'two-route strategy' from the material-concrete to the theoretical-abstract, and from the abstract to the concrete, there is not much else.[4] In this situation the introduction of a 'realist' approach (Sayer 1992 (1st ed, 1984)) was extremely helpful. First, the distinction of realism between abstract and concrete research enables the opposition between nomothetic and idiographic approaches to be transcended (Asheim and Haraldsen 1991); second, it elucidates the relation between the levels of abstraction in Marx' political economy in a non-reductionist way by explicitly stating that one strata (in the stratification of the world) cannot be reduced to the next as well as emphasizing that 'concrete intensive research' is one specific type of research; and third, it solves the problem of which level of abstraction space can be theorized as 'concrete research' is the level where space – as a property of an object and, thus, analytically inseparable from the object as such – represents an explanatory factor. Sayer underlines that 'even though concrete studies may not be interested in spatial form per se, it must be taken into account if the contingencies of the concrete and the differences they make to outcomes are to be understood' (Sayer 1992: 150). This is consistent with an understanding of geographical analyses as contextual, as well as with positioning geography as basically a synthetic discipline. According to Sayer, 'the "fetishization of space" consists in attributing to "pure space" what is due to causal powers of the particular objects constituting it. In reaction to this, some proponents of the relative concept of space have made the converse mistake of supposing that space is wholly reducible to the constituent objects, whereupon it becomes impossible to see how space make a difference, in any sense' (Sayer 1992: 148).[5]

The consequences for (economic) geography of some (Marxist) geographers reducing space to its constituent objects was also raised by Doreen Massey, who – based on her empirical analyses of the regional consequences of industrial restructuring (Massey 1984) reflected upon the radical critique of the 1970s – and argued that '"geography" was underestimated; it was underestimated as distance, and it was underestimated in terms of local variation and uniqueness' (Massey 1985: 12). This and similar reactions promoted what was called the 'new' regional geography approach, which, in my mind, came very close to solving the problems of geography basically being a synthetic discipline ('regional geography') but with the same theoretical ambitions as other social sciences ('new'), by applying a realist approach of combining abstract and concrete types of research producing theoretical informed case studies as contextual analyses providing causal explanations through retroduction.

The intermediate period – from Marxist economic geography to studies of industrial districts and regional clusters

Along with the increased attention within economic geography on the importance of contingencies, regions and local variations the global economy also underwent dramatic changes as a result of the transition from Fordism to post-Fordism (Piore and Sabel 1984). This transition led to a (re)focus on the importance of agglomerations of networked small and medium-sized firms based on a flexible production system through vertical disintegration – producing specialized, customized and semi-customized products replacing the standardized mass production of vertical integrated large firms of the Fordist period. These structural changes in the world economy, which (for some observers) paradoxically took place along with an intensified globalization, were partly caused by technological development introducing numerical operated production technology which increased the productivity of diversified batch production by minimizing the re-adjustment time of machinery, and partly by a development on the .75 of the consumer market of the western world with increased buying power, more and more demanding non-standardized products which the networked and flexible production systems of the industrial districts were able to satisfy. Thus, as can be seen, this new development is all about contingencies: technology, market trends, consumer preferences, which all takes place within the context of a capitalist economic system (or mode of production). Moreover, the new and growing role of networking, cooperation and collaboration between SMEs in industrial districts and other types of agglomerated clusters highlights the importance of non-economic factors (i.e. culture, norms, and institutions) – building social capital – for the endogenous based, economic performance of regions. Furthermore, the renewed focus on agglomerations and the regional context also provides substantiation for Porter's claim that competitive advantage is based on the exploitation of unique resources and competencies (Porter 1990), and points to economic development as a territorial embedded process, maintaining that 'competitive advantage is created and sustained through a highly localized process' (Porter 1990: 19). The continuous success of many of these new economic (regional) spaces (some of them were in fact not that new [e.g. the industrial districts of the Third Italy]) also demonstrated beyond any doubt that geography (understood as 'context' and not primarily 'distance'), contingencies and contexts still matters in a globalizing economy. It could even be argued that this tendency towards spatial concentration has become more marked over time, not less.

My own interests in studying industrial districts as a paradigmatic example of post-Fordist new economic spaces started in the early 1980s after my move to the geography department at the University of Oslo as associate professor in economic geography in August 1981. After a stay in Rome in the turn of the year 1983/84, where I travelled around in the Third Italy and among other researchers met with professors Garofoli in Pavia (now in Varese) and the late Brusco in Modena. This was the start of years of cooperation that for my own part resulted in many research projects on industrial districts, both theoretical and empirical with a focus on comparative analyses of industrial districts in Italy and the Nordic countries, as well as of districts within the Nordic countries (Asheim 1992, 1994). The theoretical work emphasized the development of a concise conceptualization of industrial districts to obtain a specific definition that distinguished districts from other forms of territorial agglomeration such as clusters and growth poles (Asheim 2000).[6]

The empirical analyses soon turned my interest towards the innovative capacity of industrial districts. Studies have shown that (firms in) industrial districts can generate incremental innovations. However, in a globalizing economy it is rather doubtful whether incremental innovations will be sufficient to avoid lock-in tendencies and promote a shift to new technological trajectories to secure the competitive advantage of firms in the districts. In this context it is necessary to keep in mind that the original rationale of industrial districts was the creation of external economies of scale to provide a competitive alternative to internal economies of scale of big companies. External economies concern the productivity of the single firm and the efficiency of the production system, obtained through an external, technical division of labour in a system of firms. Thus, it was cost or locational efficiency and not innovative capacity that is (was) the competitive (or rather comparative) advantage of industrial districts. One of the constraining structural factors in such a production system with respect to its innovative capacity (i.e. moving beyond incremental innovations) is the fierce competition between a large number of small subcontractors specializing in the same products or phases of production, and vertically linked to the commissioning firms. This promotes cost efficiency but do not represent a very innovative milieu, especially since most of these small firms are capacity subcontractors and not specialized suppliers (Asheim 1996, 2000).

All the way from Marshall's writing on industrial districts, it has been assumed that business interactions between client firms and subcontractors (exploiting localization economies) and knowledge flows were co-occurring (and co-located) phenomena. Furthermore, it has been maintained that local interactions and collective learning processes, or what is often called 'local buzz', largely happens by just 'being there' (Bathelt et al. 2004). This might well have been the case in traditional industrial districts where tacit knowledge dominated and was diffused through the industrial atmosphere ('in the air') created by the 'fusion' of the economy and society (Piore and Sabel), for example, by informal networks run by trust and civic society based social capital. However, in a contemporary situation where codified knowledge is becoming more important, and where transnational corporations (TNC) as well as large(r) local firms dominate industrial districts breaking up the 'fusion' and making informal networks formal, this is probably not any longer the case. Lately it has been shown empirically that there exist an uneven distribution of knowledge and selective inter-firm learning due to the heterogeneity of firms' competence bases, which effects the absorptive capacity of firms as well as diffusion capacity of districts (or clusters) (Giuliani and Bell 2005).

More than ten years ago, in my own empirical studies of Nordic and Italian industrial districts, I observed differences in the innovative capacity between districts. While Jæren, south of Stavanger in Norway, has consistently during many years demonstrated a rather impressive innovative capacity (including generating radical innovations), especially in the area of robot technology, Gnosjö in Småland in Sweden as well as the majority of traditional industrial districts in the third Italy showed low capacity for anything beyond incremental innovations. These differences were clearly related to the competence bases of the firms. The higher competence level (especially engineering skills) in the Jæren firms resulted in a higher absorptive capacity enabling cooperation with universities nationally and internationally as well as with demanding customers at home and abroad. The same situation can be found in the engineering industry in Emilia-Romagna with luxury car manufacturers in Modena, packaging industry in Bologna, and ceramic tile industry in Sassuolo (Asheim 1994).[7]

The present period – studies of regional innovation systems and learning regions

This focus on innovation turned my attention towards mechanisms for upgrading the innovative capacity of SMEs and industrial districts/regional clusters. The ideas of regional innovation systems and learning regions starting turning up around the mid 1990s. Regional innovation systems (RIS) are defined as 'interacting knowledge generation and exploitation subsystems linked to global, national and other regional subsystems' (Cooke 2004: 3). An RIS is not identical with a cluster since RIS normally supports more than one cluster. Recent work on innovation systems indicates that the region is a key level at which innovative capacity is shaped and economic processes coordinated and governed. This has among other things led to governments and agencies at various geographical levels looking at regional innovation systems as key elements of their innovation policy.

My own studies of regional innovation systems were initiated when I (in addition to being professor in human geography in Oslo [since 1993]) was associated with the STEP group in Oslo as a senior researcher and scientific advisor.[8] Here I – together with my first doctoral student (now professor), Arne Isaksen – built up research on regional innovation systems, clusters and innovation policy towards SMEs resulting in many large national and international research projects (e.g. see Asheim and Isaksen 1997; Asheim et al. 2003). This research continued when moving my chair in human geography in 1999 to a new *Centre for technology, innovation and culture* at the University of Oslo, initiated by among others Jan Fagerberg (Fagerberg et al. 2005), and finally when taking up the chair in economic geography at Lund University (after Gunnar Törnqvist) in August 2001, where a comprehensive Nordic project on SMEs and RIS was carried out 2002–3 (Asheim and Coenen 2005).[9]

This research has been further stimulated by the establishment of *CIRCLE (Centre for Innovation, Research and Competence in the Learning Economy)*, where my research group undertakes international comparisons of regional innovation systems with the aim of contributing to theoretical advances and presenting new empirical findings.[10] The research has developed and implemented in concrete studies a new approach for doing comparative analyses using the following main dimensions: industrial knowledge bases, distinguishing between industries based on analytical (e.g. biotech), synthetic (e.g. mechanical engineering), and symbolic (film industry) knowledge bases, and institutional frameworks applying the varieties of capitalism distinction between coordinated and liberal market economies (using regions in the Nordic countries and Canada as cases) (Hall and Soskice 2001). Bringing these analytical dimensions together has renewed the study of RIS, and has brought about a better understanding of the workings and impacts of RIS. I believe that innovation processes of firms are strongly shaped by their specific knowledge base, and, thus, need different competencies as well as supporting innovation policies (Asheim and Coenen 2005; Asheim and Gertler 2005). After years of influential research on the importance of territorial agglomerations for regional economic growth more work is now needed to disclose and reveal the contingencies, particularities and specificities of the various contexts and environments where knowledge creation, innovation and entrepreneurship take place in order to obtain a better understanding of factors enabling or impeding these processes. Differentiating between knowledge bases and institutional frameworks represents a first attempt of such an 'unpacking strategy'.

Concluding reflections: the 'missing links' and the way forward

This chapter has among other things demonstrated the integrative and interdisciplinary potential of (economic) geography. In the Nordic context this has especially become evident in the close cooperation with other heterodox (evolutionary and institutional) economists in the area of innovation studies. Nordic innovation research – especially on innovation systems – has always been strong internationally (Edquist 1997; Lundvall 1992, see also Fagerberg et al. 2005), as is also the case with innovation research having a geographical or regional focus. In addition to my own work on regional innovation studies, my close colleagues, Anders Malmberg in Uppsala and Peter Maskell in Copenhagen, have pioneered research on regional clusters internationally (Maskell and Malmberg 1999). A reason for this beyond the general strength of Nordic innovation research, is the fact that human geography in the Nordic countries rank among the smallest social sciences. This has partly made it necessary for human geography to focus on what it is best at – in contrast to, for example, Britain where the size and strength of geography has allowed its practitioners to expand into the domains of neighbouring disciplines and nearly do 'whatever they like', sometimes with a result not very encouraging. Partly it has 'forced' geography to exploit its interdisciplinary potential as a synthetic discipline, which among other things implies that it has to apply an eclectic strategy concerning theoretical work (the integrative potential). This, however, provides an excellent platform for cooperation with other disciplines, which, for example, is demonstrated by the collaborations in CIRCLE. Belonging to a small discipline has in general made (economic) geographers rather proactive and positive to other disciplines taking up regional questions. One example of this is Porter's work on cluster, which has been received much more positively by Nordic economic geographers (see the work of Malmberg and Maskell) than by British geographers (Asheim et al. 2006; Martin and Sunley 2003). For Nordic economic geographers Porter's work has opened the eyes of many policymakers for the importance of territorial agglomerations (continuous) for the innovativeness and competitiveness of firms and regions in a globalizing economy. Without such an eye opener this would not have been possible, due to the lack of (political) influence that follows from belonging to a minor discipline. The very nice thing about this development is that it is (with a few exceptions) only economic geographers that can carry out such research, as these subjects are not taught on advanced levels for economists, something that has strongly benefited the research funding of economic geographers. Finally, the limited size of the local milieus has made it necessary to establish an international network as well as research cooperation.[11]

There is, however, one problematic aspect connected with the co-evolution of economic geography with evolutionary and institutional economics. The focus on firms' and regions' innovativeness and competitiveness has missed out everything about the 'social', as such, the focus has only been on 'development in a region' (growth in regional per capita income) and not on 'development of a region' (impact on the level of living in regions). The blame for this cannot solely be thrown at economic geography, as the cultural turn in human geography at the end of the 1980s, based on post-modernist and – structuralist approaches – substituted concerns for real social problems (the actual problem of people) with interest in the representations of such problems. Thus, neither economic geography nor cultural turn-human geography took any responsibility for studying social problems. This is a paradox when thinking about the role that the demand for 'social relevance' played in radicalizing and modernizing the discipline around the 1970s. The work on learning regions as development coalitions (Asheim 2001), inspired by action oriented organizational research, may represent a small exception and a starting point for alternative research in economic geography. Development coalitions refer to a bottom-up approach based on broad mobilization promising at one and the same time economic growth and job generation as well as social cohesion. Another approach potentially bridging the gap between the economic and the social is Florida's differentiation between business climate vs people climate (Florida 2002). So far the focus of economic geographers has solely been on clusters and RIS to improve the competitive conditions of business, while ignoring the living conditions of people. Experiences from the Nordic countries – enjoying synergy effects between efficiency and equity – shows that caring for people by strengthening (and not dismantling) the welfare state is good for employment, innovativeness, competitiveness and economic growth (Hall and Soskice 2001).[12] This has also very much to do with contextualization, theoretically as well as empirically.

Notes

1. This points to interesting aspects of Nordic business schools (i.e. in Norway, Sweden and Finland) offering economic geography as an optional subject. Many chairs in economic geography in these countries have such a background.
2. It is interesting to note that the development and increased use of quantitative techniques, which came late to human geography compared to other social sciences, was called the 'quantitative revolution' in geography and not in other disciplines. This – I think – indicates the void found within human geography of not having a social object of study that could constitute the basis for geographical theoretical work.
3. In the literature on the history of geographical thought this is often called the 'liberal' response, because it was not primarily a reaction towards positivist methodologies and methods.
4. This guideline is of course potentially highly relevant, but so general that it could as easily be interpreted as a defence for a pure deductive approach.
5. An understanding of space as a property of an object, and, thus, eliminating the distinction of the relative conception of space between the spatial and the non-spatial, was introduced already in 1973 by David Harvey with the concept relational space in his book *Social Justice and the City*, which represented his personal transition from a liberal position (part one) to a socialist (or radical) one (part two). In the introductory chapter of the book he writes that 'the view of relative space proposes that it be understood as a relationship between objects which exists only because objects exist and relate to each other (what Sayer calls the spatial relations of "between-ness" (my comment)). There is another sense in which space can be viewed as relative and I choose to call this relational space – space regarded, . . . , as being contained in objects in the sense that an object can be said to exist only in so far as it contains and represents within itself relationships to other objects' (Harvey 1973: 13). However, this position runs the risk of reducing space to its constituent objects, which Harvey actually has done by arguing for the possibilities of theorizing space in 'abstract research' as part of a theory of the space economy of capitalism (e.g. Harvey 1982).
6. Industrial districts can be understood as one type of a cluster. The specific characteristics of the (traditional, Italian) industrial district (ID) compared to Porter's

cluster is: (a) the dominance of SMEs in Ids; (b) the whole value chain located within the district; and (c) the embedding of the economy within the broader society. However, as a result of globalization processes industrial districts are becoming more like 'normal' regional clusters as a result of: (a) FDIs; (b) outsourcing of parts (the labour intensive and polluting) of the value chain to countries in Eastern Europe or the Third World; and (c) group formations within the districts (Asheim et al. 2006).

7. Also with respect to an explicit focus on innovation Porter's cluster approach differs from an industrial district approach, as Porter links competitiveness and innovativeness in making competitive advantage into a more dynamic principle requiring continual innovation for its reproduction.
8. The STEP group (Studies in innovation, technology and economic policy) is a Research Council funded, independent 'think tank', established in 1993 by Keith Smith (now professor at University of Tasmania), which today has merged with another research institute in Oslo under the name NIFU-STEP.
9. I am the third chair in economic geography in Lund. The first was the Estonian refugee, Edgar Kant from University of Tartu, who had studied under Walter Christaller and brought the knowledge of Central place theory to Sweden on arrival during the Second World War.
10. CIRCLE is a Centre of Excellence in innovation system research funded by VINNOVA (Swedish Agency for Innovation Systems) and Lund University, which I initiated. It is the largest of four such centres in Sweden, and is a cooperation between three different faculties at Lund University with Charles Edquist as the first director.
11. For my own part being one of the editors of *Economic Geography* since 2000 is one example of this international orientation, another is being published in *The Oxford Handbook of Economic Geography* (Clark et al. 2000).
12. An ongoing project founded by the European Science Foundation, which I coordinate, is currently analysing the assumptions of the creative class approach by, in a modified version, adapting to a European context. The project is called Technology, Talent and Tolerance in European Cities and counts Denmark, Sweden, Norway, Finland, the Netherlands, Germany, Switzerland and UK.

References

Abler, R., J. Adams and P. Gould (1971) *Spatial Organization: the geographer's view of the world.* Englewood Cliffs, NJ: Prentice-Hall.

Asheim, B. T. (1985) 'Capital accumulation, technological development and the spatial division of labour: a framework for analysis', *Norwegian Journal of Geography*, 39: 87–97.

Asheim, B. T. (1992) 'Flexible specialization, industrial districts and small firms: a critical appraisal', in H. Ernste and V. Meier (eds) *Regional Development and Contemporary Industrial Response: Extending Flexible Specialization*, pp. 45–63. London: Belhaven Press.

Asheim, B. T. (1994) 'Industrial districts, inter-firm co-operation and endogenous technological development: the experience of developed countries', in *Technological Dynamism in Industrial Districts: An Alternative Approach to Industrialization in Developing Countries?*, pp. 91–142. New York and Geneva: UNCTAD.

Asheim, B. T. (1996) 'Industrial districts as "learning regions": a condition for prosperity?', *European Planning Studies*, 4(4): 379–400.

Asheim, B. T. (2000) 'Industrial districts: the contributions of Marshall and beyond', in G. Clark, M. Gertler and M. Feldman (eds) *The Oxford Handbook of Economic Geography*, pp. 413–31. Oxford: Oxford University Press.

Asheim, B. T. (2001) 'Learning regions as development coalitions: partnership as governance in European workfare states?', *Concepts and Transformation: International Journal of Action Research and Organizational Renewal*, 6(1): 73–101.

Asheim, B. T. and L. Coenen (2005) 'Knowledge bases and regional innovation systems: comparing Nordic clusters', *Research Policy*, 34: 1173–90.

Asheim, B. T. and M. S. Gertler (2005) 'The geography of innovation: regional innovation systems', in J. Fagerberg, D. Mowery and R. Nelson (eds) *The Oxford Handbook of Innovation*, pp. 291–317. Oxford: Oxford University Press.

Asheim, B. T. and T. Haraldsen (1991) 'Methodological and theoretical problems in economic geography', *Norwegian Journal of Geography*, 45: 189–200.

Asheim, B. T. and A. Isaksen (1997) 'Location, agglomeration and innovation: towards regional innovation systems in Norway?', *European Planning Studies*, 5(3): 299–330.

Asheim, B. T., P. Cooke and R. Martin (eds) (2006) *Clusters and Regional Development.* London: Routledge.

Asheim, B. T., A. Isaksen, C. Nauwelaers and F. Tödtling (eds) (2003) *Regional Innovation Policy for Small-Medium Enterprises.* Cheltenham: Edward Elgar.

Bathelt, H., A. Malmberg and P. Maskell (2004) 'Clusters and knowledge: local buzz, global pipelines and the process of knowledge creation', *Progress in Human Geography*, 28: 31–56.

Clark, G., M. Feldman and M. Gertler (eds) (2000) *The Oxford Handbook of Economic Geography.* Oxford: Oxford University Press.

Cooke, P. (2004) 'Evolution of regional innovation systems: emergence, theory, challenge for action', in P. Cooke, M. Heidenreich and H. J. Braczyk (eds) *Regional Innovation Systems*, 2nd edn, pp. 1–18. London: Routledge.

Edquist, C. (ed.) (1997) *Systems of Innovation: Technologies, Institutions and Organisations.* London: Pinter.

Fagerberg, J., D. Mowery and R. Nelson (eds) (2005) *The Oxford Handbook of Innovation.* Oxford, Oxford University Press.

Florida, R. (2002) *The Rise of the Creative Class.* New York: Basic Books.

Giuliani, E. and M. Bell (2005) 'The micro-determinants of meso-level learning and innovation: evidence from a Chilean wine cluster', *Research Policy*, 34: 47–68.

Hägerstrand, T. (1974) 'Tidsgeografisk beskrivning – syfte och postulat', *The Swedish Geographical Yearbook*, 50: 86–94.

Haggett, P. (1965) *Locational Analysis in Human Geography.* London: Edward Arnold.

Hall, P. and D. Soskice (eds) (2001) *Varieties of Capitalism: The Institutional Foundations of Comparative Advantage.* Oxford: Oxford University Press.

Harvey, D. (1969) *Explanation in Geography.* London: Edward Arnold.

Harvey, D. (1973) *Social Justice and the City.* London: Edward Arnold.

Harvey, D. (1982) *The Limits to Capital.* Oxford: Blackwell.

Lundvall, B.-Å. (ed.) (1992) *National Innovation Systems: Towards a Theory of Innovation and Interactive Learning.* London: Pinter.

Martin, R. and P. Sunley (2003) 'Deconstructing clusters: chaotic concept or policy panacea?', *Journal of Economic Geography*, 3(1): 5–35.

Maskell, P. and A. Malmberg (1999) 'Localised learning and industrial competitiveness', *Cambridge Journal of Economics*, 23: 167–86.

Massey, D. (1984) *Spatial Division of Labour: Social Structures and the Geography of Production.* London: Macmillan.

Massey, D. (1985) 'New directions in space', in D. Gregory. and J. Urry (eds) *Social Relations and Spatial Structures*, pp. 9–19. London: Macmillan.

Piore, M. and C. Sabel (1984) *The Second Industrial Divide: Possibilities for Prosperity.* New York: Basic Books.

Porter, M. (1990) *The Competitive Advantage of Nations.* London: Macmillan.

Sayer, A. (1992) *Method in Social Science: A Realist Approach*, 2nd edn. London: Routledge.

Schaefer, F. K. (1953) 'Exceptionalism in geography: a methodological examination', *Annals of the Association of American Geographers*, 43: 226–49.

Smith, D. M. (1979) *Where the Grass is Greener: Living in an Unequal World.* London: Penguin.

15 Approaching research methods in economic geography

William B. Beyers

Economic geography encompasses a rich variety of research topics focused on regions, systems of regions, consumers, individual businesses, aggregations of businesses, and their trade relations. It has been my great fortune to be engaged in the practice of economic geography for over 40 years. In that time I have undertaken a wide variety of research projects of a largely applied nature that have been published in academic journals and books, and in a consulting environment.[1] In this brief chapter I would like to address some lessons I've learned in the process of undertaking this research, using my own work as a basis for these lessons.

The diversity of topics that I have addressed over the span of my career has continued to invigorate me as I've continued my work. When I first started out as an economic geographer, 'economic' was pretty much equated with manufacturing on the part of industrial geographers, and there was confusion about the difference between urban, economic, and industrial geography. Some were coming at their research in a very empirical manner, and others from style framed by the 'quantitative revolution' – which was very much associated with the University of Washington before I became a graduate student there. I came to my position with no worry about whether what I was doing was too descriptive, not rooted enough in theory, or too applied. Washington, in the wake of the quantitative revolution, was a department that was immensely practical; we adopted theory and methods needed to attack the problem at hand. I believe that one of the contributions that I have made over my career has been to be a contributor to multiple arenas of debate; those in the rarified academic world of journals, as a faculty member helping to educate and train undergraduate[2] and graduate students in the field of economic geography, and as someone actively involved in the formulation of public policy. I will argue that my impact – as a person – has been stronger due to this multiplicity of professional engagements.

Each scholar has their own unique perspective on the scope of their inquiry, their methodology and theory, and the type of data that they wish to bring to bear on their project. I will be perfectly up front in saying that I was trained in a Regional Science mode of inquiry that values quantitative analysis, formal models, and the use of theory to frame research methods. Most of my work has been quantitative and inductive in nature, using primary and secondary data sources. However, my work has also depended upon qualitative sources, and in many cases has been rooted in a concern with history or change. I have organized this essay around eight points that I am illustrating with my own research; space limitations do not allow me to present the findings from the research I am citing in the form of tables. Instead, it is my intention to critically comment on this work in the context of these eight points. Here goes!

The research question(s) frame the methods needed

Most research projects that I have undertaken have been motivated by either my own curiosity or by an outside request. In both cases the research questions have generally been 'on the table', and have defined how I have approached my research. In some cases my research has been stimulated by the work of others, and I have chosen to push the envelope on that work, and to develop these themes further. I will use three papers to illustrate this approach.

Washington State was a pioneer in the development of survey-based input–output tables, producing the first table benchmarked against the year 1963, as well as tables for 1967, 1972, and 1982. One of my first papers used methods developed by Leontief and Carter to analyse structural change in the 1963 and 1967 Washington input–output models (Carter 1970; Leontief 1953). This approach required standardizing the definitions of the sectors for both years, then calculating inverse matrices, and then properly multiplying final demand vectors to estimate output. The mathematics involved was exactly as developed by Leontief and Carter, although we did not have the resources to engage in the types of price standardization that characterized their research. At the time that this research was undertaken, we did not have evidence regarding the stability of multipliers in regional input–output models. The paper found that the regional structure was less stable than the technical requirements, but also found evidence of business cycle effects (Beyers 1972). Further research of a similar nature by Conway also found business cycle effects, and helped make the case for periodic re-measurement of regional input–output relations (Conway Jr 1977). Many years later I revisited this topic, with data available over a much longer time period, using data from nonsurvey updates of the Washington input–output models, and I found that changes in regional interindustry structure had been modest, even though the shares of output of various sectors had been dramatically altered (Beyers 2001).

Another paper involving input–output models made use of existing models and research methods developed by others to focus on the empirical identification of key sectors. However, in this chapter I innovated the use of input–output multipliers for the analysis of forward linkages using an inverse matrix based on sales coefficients, and the use of purchases coefficients to derive multipliers related to backward linkages (Beyers 1976). The paper also cast these measurements into an interregional model environment, and showed how change in geographic scale influenced the identification of key sectors.

A third paper (with David Lindahl) used Michael Porter's definitions of competitive strategies to classify responses of a set of producer service establishments to certain questions regarding firm's perceptions of the bases of their competitive edge. We then tested the performance of these firms in terms of sales per employee and growth in sales, to ascertain which of these competitive strategies were superior for producer service establishments (Lindahl and Beyers 1999). In this paper we tried to be faithful to Porter's definitions of competitive strategies, but we also used discriminant analysis to demonstrate that there were other viable strategies being employed by producer services beyond those defined by Porter.

Each of these papers were driven primarily by methods or models developed by others, and the primary goal was to provide evidence in a different environment of their robustness.

Most inquiries, but not all, are driven by clear research questions

While most research projects start with a fairly clear research question, and the methodology is also clear as to how to approach these research questions, in many cases the research is exploratory, and the approach needs to be developed as the project proceeds. A good example of work of this type that I was involved with is a paper that resulted from the synthesis of various strands of data gathered in a large NSF-funded project focused on the producer services. My co-author David Lindahl and I realized that we could possibly classify responses of the firms involved in this project into a taxonomy that would allow us to characterize their development sequences, and possibly provide a test of some business strategy literature, and Taylor and Thrift's model of segmented industries (Ansoff 1965; Taylor and Thrift 1983). Lindahl and I experimented with various combinations of variables that were included in our database, recording our classification in colored chalk on about 20 feet of blackboard, and discussing among ourselves the positioning of individual businesses in this classification scheme. This exercise involved 418 detailed questionnaires with a mixture of qualitative and quantitative information, and ultimately yielded a test of the Ansoff model that showed that by being adaptive firms were rewarded with growth. It also led to a classification that documented a wide variety of adaptive behavior, ranging from firms that were failing to those that were soaring in sales growth to those that were just stagnant (Beyers and Lindahl 1997). This classification turned out to have some similarities to that developed by Taylor and Thrift, but had its own distinctive structure. This paper was not visualized when we began this research project, and it was only after we began to study patterns of responses to multiple questions that it occurred to us that we could develop these classifications of firm behavior.

In other cases my work has been purposefully exploratory. We have little in the way of interregional trade data in the United States, and while we have many models of regional economies, we do not have a rich legacy of multi-regional models. I became interested in taking the bits and scraps of data on interregional trade that came from regional input–output models, and tried to speculate about possible interregional structures. I first developed a hypothetical interregional interindustry matrix, with a specialized industry in each region, and a generic local services sector (Beyers 1978). I am not aware of anyone else attempting this kind of simulation, but it was an important effort to make, given how 'open' regional economies are – that is to say their trade relationships are typically much stronger with other regions than internally. This model was set into a system of equations that produced interregional income flows and interregional final demands, such that the output and income distribution among the regions evolved over time. I recall presenting this paper at the regional science meetings in Krakow, Poland, and had lots of computer printout containing the tables from this speculative modeling that attracted considerable suspicion from border guards who were sure I was out to sell all those numbers! I followed up this model with one that had a multi-regional demographic accounting model integrated with it, and explored the evolution of populations and economic activity over time with this system (Beyers 1980).

I further developed models of this nature after obtaining some support from National Science Foundation (NSF) to explore spatial linkage patterns of businesses located in Washington State with their markets and sources of supply elsewhere in the United States. This work found that the interaction among the states showed a gravity-model like pattern for Washington firms with clients and suppliers located in other parts of the United States, and tied the levels of activity in each region into estimates developed by Polenske in relation to the Multiregional Input Output (MRIO) model (Polenske 1970). The models were configured to have most of the interindustry multiplier effects be interregional (rather than intraregional), in accord with the data from regional input–output accounts (Beyers 1974). This kind of modeling is in many ways dreaming with numbers, but the general properties of the results appeared not to be counterintuitive. I also experimented with the use of drawings to illustrate possible alternative spatial linkage configurations, as opposed to using a gravity model in each region (Beyers 1981). This work also involved the use of cluster analysis to decompose the data in the national input–output model into broad categories of linkages. It is unfortunate that statistical accounts in the United States have failed to represent more realistically trade relationships among regions.

Frequently theory or models underlay the research approach, and motivate the type of data sought

In contrast to the above point, where off-the shelf data were used to undertake a particular type of analysis, it has also been common in my work to have a model form the underpinning for a particular piece of analysis, and to then go gather data to be used with this model. I've undertaken many economic impact studies that are structured in this way, most of them using the Washington State input–output model, or a reduced form of it for a sub-state region. A good example of this type of work are the economic impact studies undertaken for ArtsFund, an organization in Seattle that collects from corporate donors funds that are passed to non-profit arts organizations in our region. ArtsFund has sponsored three economic impact studies for King County arts organizations, based on the years 1992, 1997, and 2003. Each of these studies has involved extensive survey research, including a major survey of patrons and a survey of arts organizations (Beyers and GMA Research Corporation 2004). Patrons were asked about spending in relation to their arts experience, but were also asked a number of other questions, including open-ended qualitative questions regarding the role of the arts in the community and to them personally. My role has been to help design these studies, and to do the numerical analysis of the results of the surveys.

In the case of these economic impact studies there is a clear model that is being 'fit' through the gathering and use of particular data. However, in other cases one has a sense of the 'model' that you are seeking to fit data to, but cannot be sure about exactly how to represent the 'model'. For example, one of the goals of the producer service project referred to above was to evaluate the flexibility thesis in the context of the producer services (Christopherson 1989; Gertler 1988). This 'model' of flexible production was not only debated, it was a 'soft' concept compared to the input–output model that is a set of linear equations. We explored various facets of the flexibility issue in our project, including changes in the mix of full-time, part-time, and contractual workers (we found a modest increase in contingent work), the way in which new jobs were approached, the use of outside specialists, collaboration, and the evolution of what services were offered (Beyers and Lindahl 1999). In this work we had in the back of our minds the flexibility model that was in the popular literature, but were providing a test of it guided by the particularities of our own research agenda. The point here is this: there are many different types of models, which range from very precisely defined mathematical systems to general frameworks that have some orderly properties, but are not codified with rigid structures. In our research we need to be embracing these different frameworks with data appropriate to the type of model we are developing.

Primary data gathering is frequently needed, but many projects can be undertaken entirely with secondary data

While projects of the type discussed in the preceding section involve primary data gathering in order to accomplish their purposes, not every project needs such information. My research has often been involved with data that come entirely from secondary sources. The wealth of statistical information that is at our fingertips today on the Internet is a far cry from the statistical environment we were in some decades ago. One of my first forays into the use of secondary data for national scale analysis was the result of a request from Brian Berry. When he was President of the Association of American Geographers (AAG) he held several sessions focused on trends in the economy, and he asked me to analyse some data on trends in regional economies in the United States. Berry was involved in the conceptualization of the Bureau of Economic Analysis (BEA) Economic Areas concept, and as BEA began to provide regional data back in the 1970s, the medium was not in data files as we now know them, but rather paper printouts of special tabulations. The BEA Economic Areas are a regionalization of the United States economy that aggregates the approximately 3141 counties into about 175 metropolitan-area focused 'core' areas, surrounded by nonmetropolitan 'peripheral' areas. I received a set of data for these regions for the 1965–75 time period, and spent months coding them onto punch-cards so that I could produce analyses of trends in income among the BEA economic areas (Beyers 1979).

This particular assignment actually turned out to be a pivotal moment for my research career for two reasons. First, I discovered how vibrant the service economy was and how strongly it was associated with regional trends. Second, I realized that non-earnings income (transfer payments and dividends, royalties and rents) were growing rapidly as sources of personal income. Geographers had not addressed the role of the latter, in part because regional data were only now becoming available about these components of the personal income stream. In the years since undertaking this project, I have repeatedly used the BEA economic area regionalization to track trends in the United States economy, and have recently used these data in the context of a minimum requirements model to argue that all regional growth in the United States in recent years can be explained by trade in services (Beyers 2005). There are fewer analyses of this type than there should be, in part because of problems with the disclosure laws that pose difficulties when aggregating data from the county level to the level of the BEA regionalization. These difficulties have thwarted some from undertaking national scale analyses, as have changes in counting methods (e.g. the shift from the Standard Industrial Classification (SIC) to the NAICS classification systems).

While many projects can successfully be undertaken with secondary data, it is also common for there to be a mixture of primary and secondary data use to make arguments. An example of work of this type is my recent focus on cultural industries (Beyers 2002). This work was a response to a request for a presentation to RESER, the European Service Industries Research Network in Bergen, Norway. While I had done work on arts and cultural organizations as described above, I had not previously focused nationally on the cultural industries scene. In this paper I mixed together a variety of types of data, ranging from analyses of the personal consumption expenditures accounts that showed rising demand for spending on cultural services (in real $), to data from studies I had undertaken of recreation, arts, and sports. I tried to use these data to contextualize the relative importance of these activities in the national economy, and used BEA data to try to identify something about the geography of consumption of these activities. I reported data on the structure of income and expenditures, as well as regarding the unequal incomes earned by professional sports figures and people working in the arts. I also brought various results from economic impact studies together to show the relative contribution of components of these sectors to the regional economic base. I think that this hybrid approach worked well to touch upon a number of key attributes of a relatively understudied part of our economy.

IT has allowed a gradual expansion of the power of our research relative to its cost, and has definitely had an impact on the dissemination of research

Information technologies (IT) have had a revolutionary impact upon our computational capabilities over the course of my career. The development of computing power, software, and the Internet has all played a role in advancing our ability to ask difficult questions and undertake complex analyses. These changes have also changed the way we can share our results – instantaneously – to the planet, as Thomas Friedman has argued cogently (Friedman 2005). IT has not changed the need for scholars to conceptualize important research questions, seek funding to answer them, and to do the hard work of data gathering that is necessary to be able to answer those research questions. However, it has given us an ability to process and display results more quickly, and to engage in larger scale numerical analyses. It has reduced the labor required to undertake many projects, and has allowed a greater level of involvement of students in the classroom with analyses not possible decades ago.

The availability of data on the Internet has been expanding dramatically, and this will further help economic geographers to engage in more sophisticated analyses, especially as it becomes easier to import data into a cartographic environment. The kinds of numerical analyses that I described above were incredibly time consuming in the old world of punch-cards and Fortran programming, and today I am sure that I could do in Excel in a few hours what it took me days to do in the 1970s. While it is great that IT has made our lives easier, there is a danger of having analysis driven by what it is easy to do with modern computing systems, rather than standing back and making sure that we are asking the right questions, and gathering the data to answer them.

Results are not always what you expect, but in such cases new insights are frequently generated that advance our knowledge base

One of the wonderful experiences in economic geographic research is that you encounter findings that are not what you thought you were going to obtain. My discovery of the importance of non-earnings income described above is a case in point, and many years later my former student Peter Nelson and I codified this into an extended economic base model (Nelson and Beyers 1998). This was followed by another student, Andy Wenzl, being clever enough to implement the model that Peter Nelson and I conceptualized, and to show that county income structure was systematically related to differences in size in Washington State (Wenzl 2003).

When I undertook a large NSF project in the mid-1990s, by chance I was awarded some additional funds to do many more interviews in rural America. One of our findings from this research – and this was not anticipated – was that there was a cohort of producer service firms out in rural America that were not dependent upon local markets. They sold almost all of their services someplace else. A fair chunk of these businesses were proprietors, and following a term coined by the now-defunct Center for the New West in Denver, we labeled these people 'Lone Eagles'. Alongside them were businesses with employees who were also found out in the rural West, and we labeled these firms 'High Fliers'. David Lindahl and I wrote a paper about these firms, that we had no idea would be uncovered in this research project, and there is only one project in my entire career that has led to more e-mails and telephone calls (a study of the Mariners discussed in the next section) (Beyers and Lindahl 1996). I continue to have telephone calls from all over the United States from people observing the same phenomenon – small firms with nonlocal markets, and in Europe it has been recognized that entrepreneurs of this type are also important. Research of this type has challenged (unexpectedly) long-held biases about the power of the 'world cities', and Fortune-500 corporations. This is not to say that they are unimportant, but rather it is to say that there are other factors operating on the economic landscape as well. We would never have discovered our Lone Eagles or High Fliers without costly survey research, and it was only serendipitous that we received the funds to do these interviews.

It is critically important to engage in research that has value to the applied research community (including in a service capacity), as well as to be expanding our basic research understanding

I have enjoyed having one foot in the basic research community, and one foot in the applied research community, but also a third foot in the community in a service capacity. Readers who follow the link to my website will find lists of publications, and also a long list of consulting reports. Not listed on this website are my efforts in the community as a member of a task force, board, or committee, where my expertise in the academic side of my research has been extended into the world of community service. It may be my department or this region, but there has always been a strong pull to work with the community on research projects, and these have informed my scholarly research and classroom teaching. These types of community involvement have continuously led me to have types of knowledge and understanding that I have used in the classroom, and in my publications, as in my 2002 paper on cultural services (Beyers 2002).

A good example of this is the work that I've done with Dick Conway (a local economic researcher) as a consultant on the economic impact of the Mariner's baseball team. The Mariner's have threatened to leave Seattle several times, and these threats precipitated King County (where Seattle is located) to hire us to do economic impact studies of the team and its fans spending (Conway and Beyers 1994). We've done this several times, but a key point is that this analysis has been focused upon by people in many other regions, by students, by writers, and people are constantly calling to see if we've done an update, as they have hunger for information about the economic impact of baseball. We have not done an update since 1994. However, the point here is that doing work of this type leads to public interest in your work, and as economic geographers we need to keep a sharp eye on being in the press. Several years ago the MacNeil-Lehrer NewsHour came to interview me about this study, at a time when there was a strike of the major league players. After taping their show in my office, I got a call just before they were to air my ten second sound bite worrying about labeling me as a geographer, as they said that would confuse their audience. So, after a long discussion, we agreed that I'd be labeled 'economic geographer'. It would be great if more of us were engaged with our communities, so that the press would naturally turn to us for information on the regional economy.

We have an ongoing responsibility to transfer knowledge of research methods to our students, and assure that their development will expand the importance of economic geographic understanding in the community of scholars

I have been fortunate to be associated with a department that has a strong economic geographic tradition. In this respect, it has been important to me to pass down to my students the passion to be involved with this field, and to work with them in a research environment. At Washington we have an active program of undergraduate research, including hiring undergraduate students to be involved with faculty research. We have also been very fortunate in having wonderful graduate students, who have pushed their professors into collaborative relationships. I've mentioned several students in this chapter, and would like to end by illustrating this argument through one such association, with Peter B. Nelson, who is now on the faculty of Middlebury College in Vermont.

Peter Nelson came to Washington for graduate studies from the wonderful undergraduate program in geography at Dartmouth. He and I ended up working on a field-based project in a set of rapidly growing communities in the rural West, and after our days of interviewing, we often sat in our motel room in the evening writing up our day's experiences on our laptops (with some drinks). We had a framework for these rural interviews, after we studied our results we found that there were a number of features of the interviews that we had done that were not what we expected. We talked about this as he developed his dissertation research proposal, and worked together in putting together a paper that captured some of these unexpected findings (Beyers and Nelson 2000). The point here is that I was not dominating this faculty-student relationship – it was naturally collaborative. And after Pete finished his degree, we have had continued collaboration. This is crucial for economic geographers, in some measure due to the variety of modes of research that we engage in.

A brief concluding remark

Economic geography is an exciting field, and there are many approaches to the subject. The key point I've tried to make here, by way of reference to my own work, is that there is no one methodological or philosophical perspective that works for each person. Each reader will construct for themselves their own approach to their research. Each contributor to this book has their own bag of tricks. What has been of importance for me over my career has been the use of multiple-methods, and a strong engagement with both primary data and formal models. At the same time, it has also been very important to be involved with projects that are in demand in the community, and to invigorate students with a strong interest in undertaking economic geographic research. I hope that economic geographers who read this account of my work will agree with my multi-faceted thrust to the field. And, I know that they will continue to stake out their own approach to the field.

Notes

1. Please go to http://faculty.washington.edu/beyers and find a link to my vita for a compilation of publications.
2. I would guess that I have lectured over 7,500 students in this time period: I am having the children of former students in my classes these days!

References

Ansoff, H. I. (1965) *Corporate Strategy: An Analytic Approach to Business Policy for Growth and Expansion*. New York: McGraw Hill.

Beyers, W. (1972) 'On the stability of regional interindustry models: the Washington data for 1963 and 1967', *Journal of Regional Science*, 12(3): 363–74.

Beyers, W. (1974) 'On geographical properties of growth center linkage systems', *Economic Geography*, 50: 203–18.

Beyers, W. (1976) 'Empirical identification of key sectors: some further evidence', *Environment and Planning A*, 8: 231–6.

Beyers, W. (1978) 'On the structure and development of multiregional economic systems', *Papers of the Regional Science Association*, 40: 109–33.

Beyers, W. (1979) 'Contemporary trends in the regional economic development of the United States', *Professional Geographer*, 31(1): 34–44.

Beyers, W. (1980) 'Migration and the development of multiregional economic systems', *Economic Geography*, 56: 320–34.

Beyers, W. (1981) 'Alternative spatial linkage structures in multiregional economic systems', in J. Rees, G. J. D. Hewings and H. A. Stafford (eds) *Industrial Location & Regional Systems*. Brooklyn, NY: J. F. Bergin.

Beyers, W. (2001) 'Changes in the structure of the Washington state economy, 1963–1987: an investigation of the patterns of inputs and the mix of outputs', in E. Dietzenbacher and M. Lahr (eds) *Input-Output Analysis: frontiers and extensions*, pp. 100–20. London: Macmillan Press Ltd.

Beyers, W. (2002) 'Culture, services, and regional development', *Service Industries Journal*, 22(1): 4–34.

Beyers, W. (2005) 'Services and the changing economic base of regions in the United States', *Service Industries Journal*, 25(4): 461–76.

Beyers, W. and GMA Research Corporation (2004) *Economic Impact of Arts and Cultural Organizations in King County 2003*, report for ArtsFund, a Seattle nonprofit arts support organization.

Beyers, W. and D. P. Lindahl (1996) 'Lone eagles and high fliers in rural producer services', *Rural Development Perspectives*, 12(3): 2–10.

Beyers, W. and D. P. Lindahl (1997) 'Strategic behavior and development sequences in producer service businesses', *Environment and Planning A*, 29: 887–92.

Beyers, W. and D. P. Lindahl (1999) 'Workplace flexibilities in the producer services', *The Service Industries Journal*, 19(1): 35–60.

Beyers, W. and P. Nelson (2000) 'Contemporary development forces in the nonmetropolitan west: new insights from rapidly growing communities', *Journal of Rural Studies*, 16: 459–74.

Carter, A. (1970) *Structural Change in the American Economy*. Cambridge: Harvard University Press.

Christopherson, S. (1989) 'Flexibility in the US service economy and the emerging spatial division of labor', *Transactions, Institute of British Geographers*, 14: 131–43.

Conway Jr, R. S. (1977) 'The stability of regional input-output multipliers', *Environment and Planning A*, 9: 197–214.

Conway Jr, R. S. and W. B. Beyers (1994) *Seattle Mariners Baseball Club Economic Impact*, report prepared for King County, Dick Conway & Associates and Department of Geography, University of Washington, Seattle.

Friedman, T. L. (2005) *The World is Flat: A Brief History of the Twenty-First Century*. New York: Farrar, Straus and Giroux.

Gertler, M. (1988) 'The limits to flexibility: comments on the post-fordist vision of production and its geography', *Transactions of the Institute of British Geographers*, 13: 419–32.

Leontief, W. (1953) *Studies in the Structure of the American Economy*. New York: Oxford University Press.

Lindahl, D. P. and W. B. Beyers (1999) 'The creation of competitive advantage by producer service firms', *Economic Geography*, 75: 1–20.

Nelson, P. B. and W. B. Beyers (1998) 'The economic base model in new clothes: responding to structural trends in the rural west in the 1990s', *Growth and Change*, 29: 321–44.

Polenske, K. R. (1970) *A Multiregional Input-Output Model for the United States*. Cambridge: Harvard Economic Research Project.

Taylor, M. J. and N. Thrift (1983) 'Business organization segmentation and location', *Regional Studies*, 17: 445–65.

Wenzl, A. J. (2003) 'Consumption side up: the importance of non-earnings income as a new economic base in rural Washington state', unpublished M.A. Thesis, University of Washington.

16 Manufacturing, corporate dynamics, and regional economic change

H. Doug Watts

Introduction

It can be argued that one of the central aims of economic geography is to describe changes in regional economic structures and to understand why such changes takes place. This is most frequently explored in terms of jobs gained and lost rather than in terms of output. A focus on jobs rather than output arises from both the ready availability of regional employment data and a desire to link economic geography with public concerns about geographical variations in job opportunities. Within the wider concern with regional economic change, a particularly important group of studies focus upon the role of large multi-regional firms in guiding the geographies of the manufacturing sector. This reflects a fascination with the ways in which such firms shape the economic landscape

Over the last 50 years both theoretical and empirical investigations into the geographies of large multi-regional firms have become more sophisticated. Theoretically, the strong neo-classical economic approaches of the early part of the period have been complemented by analyses from the political economy and institutional economics viewpoints to which have been added, over the past decade, the 'cultural turn'. Space limitations restrict the discussion of empirical work mainly to studies of changing employment patterns and the factors which influence them rather than the ways in which each establishment/region can be linked by traded and untraded interdependencies into local, regional, national and international systems.

This chapter is in four parts. The first sets the context for this review and takes a broad look at changes in manufacturing and their effects on research into corporate dynamics and regional change. The second explores the changing theoretical perspectives of the last 50 years. The third looks specifically at attempts to understand patterns of employment change within large multi-regional manufacturing firms. This draws upon both empirical studies and new theoretical perspectives. The fourth and final section explores new avenues for research and the ways in which our research can feed into the policy community. The overall aim of the chapter is to reflect upon the literature within economic geography that has examined the spatial organisation of production within multi-regional firms. It is a personal reflection on over 30 years of published research in the area varying from an exploration of oligopolistic behaviour in the United Kingdom sugar beet processing industry (Watts 1971) to the European wide restructuring of production by United States and European multinationals (Watts 2003).

The manufacturing sector

In many regions in advanced economies the most striking change in economic structures has been the decline of employment in manufacturing activities in both absolute and relative terms. In the UK, for example, employment in manufacturing fell from 8.6 million in 1965 to 3.7 million today. This loss of almost five million jobs saw manufacturing's share of total United Kingdom employment fall from 37 per cent to 17 per cent. This change in economic structure is reflected in the nature of empirical work in economic geography where the proportion of work focused exclusively on manufacturing has become less important.

Despite the decline of employment in manufacturing, the importance of an understanding of the role of manufacturing in regional change should not be underplayed. Other measures of manufacturing indicate clearly its overall significance in the mix of economic activities. Taking the United Kingdom as an example again, manufacturing is responsible for two thirds of all exports by value and for about 75 per cent of the research undertaken by business organisations. It can be argued that despite manufacturing employment decline Cohen and Zysman's (1988) claim that 'manufacturing matters' is as relevant in the twenty-first century as in the twentieth.

The changing significance of manufacturing has been accompanied by the emergence of new industries and the decline in the importance of older activities. In the latter half of the twentieth century the fourth Kondratieff based on electronics, computers and aerospace research peaked (Hall and Preston 1988). These newer industries tend to be more knowledge based and to build upon new technologies and innovations. This has impacted upon the research agendas in that there has been a distinct move away from analysis of traditional industries towards a focus on the high technology sector. Further, within the older industrial sectors, analysis of the motor vehicle assembly sector has perhaps had an undue influence on economic geography as a whole. Concepts relevant to understanding changes in the vehicle assembly industry (*sic* Fordist) have been transferred (in some cases rather uncritically) to other manufacturing industries.

In both the newer industries, and in many of the more traditional ones, changes in the importance of manufacturing and the sectoral mix were accompanied by an increasing dominance of regional economies by large multi-regional and often multinational firms. Indeed, by the 1980s, such firms were the main form of organisation within the manufacturing sector. The rise in importance of the large firm led economic geographers to take an increasing interest in the impact of these organisations on global, regional and local economic systems. Looking back on significant economic geography texts (whether in the United States, Jones and Darkenwald 1965 or in the United Kingdom, Smith 1953) sector after sector are discussed with only limited reference to firms. Industries rather than firms were seen as responding to economic, social and political forces to create specific geographies. The basic building block – the firm – was seen as only of marginal relevance. Perhaps the main exceptions to this assertion were found in the work of some historical economic geographers who placed considerable emphasis on corporate interviews and archives (Warren 1970).

Admittedly, the significance of these large multi-regional firms in employment terms declined in the 1990s and smaller firms came to account for an increasing share of employment. This was due partly to an increase in output per person in large firms (which was not matched by smaller firms) and partly to the outsourcing of the non-core activities of large firms to smaller firms. As a result of the more important role of small firms, there was a shift in research interests within economic geography from large firms to small and medium-sized enterprises (SMEs). Whilst this may have reflected their increasing importance in employment terms, it also mirrored the fact that SMEs (and especially high technology and innovative firms) became of major policy interest for they seemed more amenable to government policy initiatives than the large firm. Further, SMEs fit rather well into debates on the emergence and continuation of industrial clusters, which became too central to many studies of regional performance in the 1990s, although the benefits of clusters seem increasingly challenged (Martin and Sunley 2003).

Despite the shift in research emphasis to the SME, the large firm continues to be seen as a key actor in the global economic system (Dicken 2003) and, in the United Kingdom, large firms (of over 250 employees) account for almost two-thirds of manufacturing turnover. The corporate dynamics of large multi-regional firms cannot be ignored in any move towards increasing our understanding regional economic change.

Theoretical perspectives

The ways in which economic geographers have approached the study of corporate dynamics in the manufacturing sector have reflected wider changes in economic geography. These wider changes are rather neatly summed up by Barnes (1999: 17) who notes a move from 'spatial science and location theory in late 1950s, behavioural theories of the firm in the late 1960s and structural Marxism . . . in the 1970s and 1980s'. To which might be added a recognition of the role of institutions in the 1990s. These four approaches to the study of large firms tend to be moulded together leading to an eclectic mix of theoretical perspectives and empirical investigations to inform our understanding of corporate dynamics and regional economic change.

Despite the successes of approaches based on spatial science and location theory (for an early example see Stafford 1960) it became increasingly evident throughout the 1960s and 1970s that economic geographers could not explain adequately what was going on in a region without an understanding of the corporate context in which many of the region's plants were set. This was recognised by

McNee (1960) and developed later, under the leadership of Morgan Thomas, into what might be called the Washington school of economic geography, of which the work of Hayter (1976) provides an early exemplar. A 'geography of enterprise' emerged with ideas and concepts, which were new to geography. Strictly this should have been termed a 'geography of the corporate enterprise' since the term 'geography of enterprise' might equally well apply to the smaller and medium sized enterprise.

This recognition of the role of the larger firm in regional systems seems a critical turning point. It was given further impetus, especially in the United Kingdom, by those who sought to critique the capitalist firm and the study of large capitalist firms became central to political economy approaches to economic geography (Massey and Meegan 1982). However, the emphases in the two approaches to the larger firm were rather different. Whereas the political economy approach focused much attention on macro economic forces driving the firm with relatively little attention to the details of the corporate response, the geography of enterprise approach tended to look in detail at the behaviour of the firm and the ways in which it chose to respond to the wider macro economic forces. Unlike a SME which could be pushed and pulled by market forces, the larger enterprise, although not immune from market forces, was able to plan its spatial configuration and to use its power to exploit the differences between places.

The concern with larger firms and their characteristics led economic geographers into the literature of industrial economics and management. Particularly important was the work of Simon (1955) who argued economic man (*sic*) might be a satisficer rather than an optimiser. This had significant implications for the understanding of patterns of manufacturing activity. In particular the idea was developed over a decade later by Pred (1967) who argued that behavioural approaches would suggest corporate geographies might well be influenced both by the knowledge available to corporate executives and their ability to use it.

Whilst these behavioural approaches could be linked back to earlier theories (for example, Weber 1929) economic geography has more recently moved sideways rather than in a cumulative manner. Recent research has been over keen on 'new turns' and has not built on what has gone before. It might even be argued as a topic became more difficult it was abandoned to be replaced by something that was more fashionable. Indeed, Clarke's (1996: 284) comment on human geography as a whole seems to be particularly pertinent, 'the discipline does not seem to be flowing in a linear progression at all; rather it appears that we are either going round in circles or perhaps bifurcating in radically opposed directions'.

This is seen clearly in the 'new economic geography' which is a fuzzy concept. To some this is the 'cultural turn' in which economic geographers look closely at issues of consumption, to others it is the 'geographical economics' of Krugman (1995) whilst to others such as Barnes and Gertler (1999) it is institutional geography. The small amount of contemporary economic geography which builds on the research of the 1960s and 1970s relates well to Krugman but it is increasingly marginalised within mainstream geography and finds its natural home in Regional Science. It might be argued that this shift also reflects a rejection of quantitative analysis in geography as a whole. In graduate schools in the UK, finding economic geographers interested in production rather than consumption can be difficult and finding economic geographers using quantitative methods is more difficult still.

The absence of a linear progression in research is a major weakness of economic geography and it is perhaps a weakness it shares with other aspects of human geography. It is a weakness that stands out in comparing introductory texts in economic geography with those of related disciplines such as economics and sociology. Contemporary economic texts are still happy to teach long established supply and demand concepts and contemporary sociology texts include work from the late nineteenth century. In contrast, modern texts on economic geography tend to pay little attention to early theoreticians. If my students are typical of the United Kingdom such writers also seem to have disappeared from the school syllabus too!! Similarly our extensive knowledge of the factors influencing the choice of location for a branch plant are not often reported in contemporary texts, despite their significance for regional development issues. This is not to argue that contemporary concerns and the 'cultural turn' are to be ignored but they should build more firmly on the rich inheritance of earlier work in economic geography

Corporate dynamics and the multi-regional firm

A useful way to assess the current state of knowledge of the corporate dynamics of a multi-regional firm is to recognise that changes in the 'spatial configuration' of its production system reflects the operation of three mechanisms: the entry of sites to the system, the exit of sites from the system and the expansion/contraction of those sites which are maintained as continuous elements within the production system. A pioneering attempt to measure the relative importance of these mechanisms is provided in Healey (1983). These mechanisms have received different degrees of attention from researchers within economic geography and it is useful to consider each in turn. Over the last few decades the focus of research has moved from entries, through exits, to the present interest in repeat investment.

Entries

Additions to the corporate system take the form branch plants and acquisitions. Acquisitions attracted little attention (an important early exception is Leigh and North 1978) but the factors influencing branch plant location decisions were widely researched in the 1950s–1970s (see, for example Keeble 1968) and became increasingly sophisticated both conceptually and in the analytical techniques applied to the data. In the United Kingdom good data sets also helped work in this area. There was a strong policy push on this research since new plants resulted in new jobs (even if displacement effects were ignored). Further, it was demonstrated quite convincingly that despite comments that locations were selected by 'pins in a map' there were sufficient regularities in the patterns produced in the establishment of branch plants to admit to an underlying logic. Even those cynics who regarded many sites as 'golf course locations' were silenced by the recognition that golf courses could be found in close proximity to most potential locations and therefore did not provide a way of discriminating between those locations.

An important conceptual advance was the recognition that different factors might apply at different scales. Access to a freeway might govern location within a town whilst the particular characteristics of a labour market might influence the selection of a town. Indeed, such was the progress in this field that Fothergill and Guy (1990: 43) were able to comment that 'the conclusions of these . . . (branch plant) . . . studies were sufficiently unambiguous and consistent that . . . little further research has had to be devoted to understanding branch openings'. It will be argued below that the dismissal of the need for further research was perhaps premature.

Exits

Despite the recession and job losses from 1980 onwards geographers were slow in turning their attention to the geography of job loss. The reasons for this were varied. Job loss was less important than job creation in policy terms. Large firms tended to be reluctant to talk about plant closures. Further the public were often taken in by the corporate 'excuse' of a fall in demand explaining the closure of a particular plant in a particular region. The company would not draw attention to the fact that despite the fall in demand it was retaining production in other sites and that the real reason for closure of a particular plant was the high land value of the site or a perception of a difficult labour force.

Conceptually a link can be made with the branch plant location/acquisition literature for those cases were a firm was selecting between plants making similar products. The locational factors which made a location attractive for a branch plant would (if negative) make a plant attractive for a closure whilst the characteristic which made a plant an attractive feature for acquisition would in reverse make a plant an attractive candidate for closure. Compared with the analysis of branch plant locations relatively few studies examined the theoretical and empirical issues arising from plant closures implemented by large multi-regional firms (Watts and Kirkham 1999; Watts and Stafford 1986). Within geography there has been little interest in the disposal of businesses, partly one suspects as in the short term disposal of a business as a going concern may have few job implications.

Continuing

Until quite recently there has been surprisingly little interest in job change in plants which neither open nor close during a particular period of time. It is surprising in the sense that studies of components of change often indicate job changes in plants that continue to exceed those in newly established plants or plants that close. There are clearly important questions as to why some plants within a corporate system increase in significance whilst others loose out. A particular interest at the present time is the concept of 'repeat investment' (Phelps and Fuller 2000). Why do some plants attract repeat investment? Why do others fail to do so?

Regional perspectives

The entries, exits and continuing plants of the large firm impact upon individual regional economies. Indeed, my own initial move from regional economic geography in the 1960s to the study of corporate dynamics in the 1970s was driven by the fact that it was not possible to understand changes in plants within my region of interest unless one recognised the way in which externally owned plants fitted within wider corporate structures. The role of such large firms in regional economies was highlighted in the 1970s, partly as a response to the growth of large firms themselves but also because it was a politically sensitive issue in Britain's Celtic fringes. Politicians in Scotland and Wales became concerned that many decisions affecting their economies were being taken in England or other foreign countries. Peripheral regions and countries of the United Kingdom began to see themselves as branch plant economies (Watts 1981). The impact of external ownership, which usually meant external control over major investment decisions, was seen as detrimental to as regional economy. Empirically the evidence was rather mixed in terms of the opening and closings of plants but there was clear evidence that external control could impact on the occupational mix of a region. High-level managerial jobs were often concentrated outside a peripheral region and, in the context of the United Kingdom, this was in the South East of the country. In addition, there was strong evidence that research and development (R&D) expenditures by business also had a southern bias.

Future directions: policy and research

Policy

The research in economic geography on large firms would seem to have at least two applications. Advising managers (or potential managers) of large firms on the ways in which variations between places can be exploited and informing policy-makers concerned with regional change. Knowledges can be diffused in embodied form by which trained economic geographers emerging from doctoral programme take up positions outside the discipline of geography. More conventionally, transfers of such knowledge take place through publication in both academic and, more importantly, practitioner literature. Sadly in the United Kingdom the latter is not encouraged as activities driven by the UK Research Assessment Exercise (RAE) place a premium on publication in the academic journals.

Advising business mangers (both actual and potential) can be begun by economic geographers working within and publishing in the field of management. Whilst a number of geographers dealing with entrepreneurship, retailing and/or logistics have moved into the management area this is less common amongst those geographers with expertise in the manufacturing activities of larger firms. Nevertheless, my doctoral students whose PhD theses concerned the behaviour of large multi-regional firms include one who is now head of MBA programmes at the University of Hertfordshire and another is a Senior Lecturer (Associate Professor) in Management at Queen's University Business School based in Belfast. Publishing in management journals should also be encouraged. Research by geographers on plant closures has appeared in journals such as *Management Decision* and the *International Journal of Manpower* and it has also been argued that geographical analysis of the nature of greenfield sites can inform research in human resource management (Richbell and Watts 2001).

Regional development agencies, whose basic aims often involve maintaining and creating jobs provide further audiences for our work especially that on branch plant location, repeat investments and plant closures. Knowledge transfer through PhD programmes play a part here too. One trend noticeable in the destinations of my doctoral students is a greater emphasis on employment in policy related areas. Whereas in the 1970s and 1980s PhD students tended to enter academic posts the 1990s have seen a greater interest in policy and its applications. Again using the examples of my own doctoral students, one is now head of regeneration in a local authority in the West Midlands of England and another is employed by Yorkshire Forward, the Regional Development Agency for part of the north of England. Certainly doctoral students with quantitative skills seem to have little difficulty in the policy job market.

Research agenda

It seems reasonable to argue for a return to a close interest in the role of the large multi-regional firms in the creation and maintenance of regional economic systems. It is a challenge that needs to be met. Any research agenda is inevitably personal but three themes seem particularly important if we are to increase our understanding of corporate dynamics, job opportunities and regional economic change.

Although unfashionable, it now seems vital that we revisit some of the older questions using our greater conceptual awareness and new methodological tools. This seems particularly important as many of the illustrations we are able to give of particular important processes at work are now 20 or 30 years old. Are the patterns we analysed then still in existence today? Are they explained by the same factors? A high priority should be placed on studies that replicate those completed some years ago. This will, of course, contribute to a further development of knowledge which as was noted earlier is particularly lacking in economic geography where we seem to be tempted to move on to new problems rather than explore older ones in greater depth.

We should also pick up on unresolved questions from the past. One example will illustrate this point. Studies of branch plant location indicated a marked distance decay effect whereby most firms established new branch plants a short distance from their existing operations and few firms established branch plants a long distance from their established operations. Why does this distance decay effect occur? Why does it vary between different origin regions? Although there have been speculations as to why this occurs, it does seem a careful and rigorous analysis is still required. Further, levels of explanation in multiple regression analyses of branch plant movements have tended to be low suggesting there is much more to learn in this field. This area too is of major significance because 'mobile' jobs remain very attractive to policymakers whose place marketing is used to lure into their regions greenfield investments by major firms.

Although we have a good basic knowledge of plant openings and closing by large firms (although as noted above these are capable of refinement) much less is known about the activity renewal at specific sites which a firm retains. The rapidly growing interest in 'repeat investment' deserves very strong encouragement. As Phelps and Fuller (2000: 225) observe 'questions regarding . . . (intracorporate) . . . competition are . . . central to an understanding of contemporary industrial restrucuturing, regional development and policy, yet to date little academic work has addressed itself explicitly to such questions'. Answers to such questions will provide further understanding of the ways in which large multi-regional firms create, maintain or destroy job opportunities in particular places.

Conclusion

In my fourth decade as a researcher into corporate dynamics and industrial change it seems reasonable to conclude that although our knowledge has advanced significantly since the early 1960s there are still exciting ways in which our research can move forward, especially if it builds on what has gone before. However, we need to recognise economic geography is not simply about the economic factors influencing regional economic change but that it also needs to consider the impact of social, cultural and political factors in influencing such change.

Acknowledgements

I am indebted to Roger Hayter and Howard Stafford for comments on a preliminary draft of this chapter and to the organisers and participants in the Association of American Geographers (AAG) symposium for encouraging me to think through some of the issues discussed here.

References

Barnes, T. (1999) 'Industrial geography, institutional economics and Innis', in T. Barnes and M. Gertler (eds) *The New Industrial Geography: Regions, Regulation and Institutions*, pp. 1–22. London: Routledge.

Barnes, T. and M. Gertler (1999) *The New Industrial Geography: Regions, Regulation and Institutions*. London: Routledge.

Clarke, D. B. (1996) 'The limits to retail capital', in N. Wrigley and M. Lowe (eds) *Retailing, Consumption and Capital*, pp. 284–301. Harlow: Longman.

Cohen, S. S. and K. Zysman (1988) *Manufacturing Matters: The Myth of the Post-Industrial Economy*. New York: Basic Books.

Dicken, P. (2003) *Global Shift: Reshaping the Global Economic Map in the 21st Century*. London: Sage.

Fothergill, S. and N. Guy (1990) *Retreat From the Regions: Corporate Change and the Closure of Factories*. London: Jessica Kingsley.

Hall, P. and P. Preston (1988) *The Carrier Wave: New Information Technology and the Geography of Innovation*, pp. 1846–2003. London: Unwin Hyman.

Hayter, R. (1976) 'Corporate strategies and industrial change in the Canadian forest product industries', *Geographical Review*, 66: 209–28.

Healey, M. J. (1983) 'Components of locational change in multiplant enterprises', *Urban Studies*, 20: 327–41.

Jones, C. F. and G. G. Darkenwald (1965) *Economic Geography*. London: Collier-Macmillan.

Keeble, D. (1968) 'Industrial decentralisation and the metropolis: the north-west London case', *Transactions of the Institute of British Geographers*, 44: 1–54.

Krugman, P. (1995) *Development, Geography and Economic Theory*. London: MIT Press.

Leigh, R. and D. North (1978) 'Regional aspects of acquisition activity in British manufacturing activity', *Regional Studies*, 12: 227–46.

Martin, R. and P. Sunley (2003) 'Deconstructing clusters: chaotic concept or policy panacea', *Journal of Economic Geography*, 3: 5–35.

Massey D. and R. Meegan (1982) *The Anatomy of Job Loss*. London: Methuen.

McNee R. (1960) 'Towards a more humanistic economic geography: the geography of enterprise', *Tijdschrift voor Economische en Social Geografie*, 51: 201–5.

Phelps, N. A. and C. Fuller (2000) 'Multinationals, intracorporate competition and regional development', *Economic Geography*, 76: 224–43.

Pred, A. (1967) 'Behaviour and location: foundations for a geographic and dynamic location theory: part I', *Lund Studies in Geography B*, 27: 1–128.

Richbell, S. M. and H. D. Watts (2001) 'Shades of green: the greenfield concept in Human Resource Management', *Employee Relations*, 23: 498–511.

Simon, H. A. (1955) 'A behavioural model of rational choice', *Quarterly Journal of Economics*, 69: 99–118.

Smith W. (1953) *An Economic Geography of Great Britain*. London: Methuen.

Stafford, H. A. (1960) 'Factors in the location of the paper-board container industry', *Economic Geography*, 36: 260–6.

Warren, K. (1970) *The British Iron and Steel Sheet Industry Since 1940*. London: Bell.

Watts, H. D. (1971) 'The location of the beet sugar industry in England and Wales, 1912–36', *Transactions of the Institute of British Geographers*, 53: 95–116.

Watts, H. D. (1981) *The Branch Plant Economy: A Study of External Control*. London: Longman.

Watts, H. D. (2003) 'Understanding cross-border plant closures in the EU: a UK perspective', in N. Phelps and P. Raines (ed.) *The New Competition for Inward Investment*, pp. 137–53. Cheltenham: Edward Elgar.
Watts, H. D. and J. Kirkham (1999) 'Plant closures by multilocational firms: a comparative perspective', *Regional Studies*, 33: 413–24.
Watts, H. D. and H. A. Stafford (1986) 'Plant closures and the multi-plant firm: some conceptual issues', *Progress in Human Geography* 10: 206–27.
Weber, A. (1929) *Theory of the Location of Industries*. Chicago: University of Chicago Press.

17 On the intersection of policy and economic geography

Selective engagement, partial acceptance, and missed opportunities

Amy K. Glasmeier

My chapter considers the role of academic geographers as policy advisers, and explores what may lie behind the absence of economic geographers in American national policy contexts. I look at a moment in history when scholars, loosely described as economic geographers, did weigh in on national economic policy issues. I discuss research practice in a policy context and then note that in the United States, economists dominate the practice of policy science because of their specific world view and epistemology. I then examine what happens when we do weigh in on policy issues and what happens to our ideas, including their use in unintended ways. I conclude with some topics that should receive geographical investigation – topics about which geographers are unusually quiet despite the incredible spatiality of such problems.

By way of introduction, I am speaking of the American context. I acknowledge that in the United Kingdom and other parts of the world, economic geographers are among the many academic advisers to policymakers. The peculiar history of American geography and its half century quest to be considered a 'science', the discipline's struggle with its positionality, the lack of exposure to geography in k-12 educational settings, and the dominance of policy debates by economists have constrained the discipline's ability to make relevant and necessary contributions to public discourse on social policy.

Other issues reduce the attractiveness of policy research to academic geographers. Many geographers are not motivated to acquire the skills required to conduct policy research. These tend to be drawn from economics and evaluation research. Geographers often find unattractive the epistemological orientation of policy debates, which are confined to or defined by a model of 'normal science' that uses statistical tools and techniques. The academy's lack of recognition of the value of policy research in a person's career further diminishes its relevance. There is a perceived difficulty in translating policy research into scholarly publications and extra effort is required to bridge the gulf between the languages of policy and academia.[1] While I note these five factors I will only address two of them directly in the remainder of this essay.

Articles like this usually contain some personal confessions. This intervention will be no different. First, I am not a degree-carrying geographer. Although I was trained by geographers such as Peter Hall and Richard Walker, and influenced by others such as Doreen Massey, David Harvey and Dick Peet, as well as my graduate student colleagues Michael Storper, Meric Gertler, Mary Beth Pudup, Susan Christopherson, Suzanne Hecht and others, my degrees from the University of California at Berkeley were in planning. In the early 1980s, geography and planning were intertwined. Today many of my former graduate school colleagues in the geography department are in planning programs even as some planners are in geography programs. The factors that led to this convergence are a good starting point in considering geographers' roles in policy debates.

History

Two concatenated experiences and the importance of key actors contributed to the emergence of a group of geographers and planners who were policy-oriented and sought to be policy-relevant at Berkeley in the early 1980s. Turning the clock back to that time, I was a member of a group of aspiring academics who came together at UCB and spent five years completing dissertations on various topics loosely linked with the subdiscipline of economic geography. How we converged on Berkeley is a separate story, but suffice it to say that while there we were influenced by issues and struggles occurring in the nation at the time.

The two previous decades of social activism around issues such as the Vietnam War, Women's rights, inner-city urban decline, and the rise of the environmental movement served as potent stimulants for the emergence of new social movements and citizen-based activism. Coincidental with, but largely distinct from those seeds of activism, was the economic crisis of the late 1970s when high interest rates, falling productivity, corporate malfeasance and internationalization of the economy led to massive job losses in basic industries. Whole regions such as the Industrial Manufacturing Belt came under siege as American firms shed millions of jobs in the wake of revived competitors such as Germany and the emergence of new competitors including Japan and the emerging Asian Newly Industrializing Economies, that were profoundly changing the industrial landscape (Harrison 1997; Harrison and Bluestone 1988; Harrison et al. 1980, 1982; Harrison and Glasmeier 1997). This period of tumult stimulated policy engagement and critique. In the early 1980s it was difficult to ignore the massive upheaval engulfing the nation. Such extreme change served to legitimize activism and encourage engagement with social issues of an immediate nature.

Our engagement was further facilitated by the presence of public scholars and academic activists who were working inside the 'conventional world' acting as role models for our own politicization. They included Bennett Harrison, Norm Glickman, Dick Walker and Ann Markusen, who were academics and activists. Especially important, people like them engaged the policy context by offering theoretically informed commentary about contemporary empirical evidence focused on major social issues of the day. Comparable actors in the United Kingdom were people like Doreen Massey, Peter Hall, Richard Meegan, and many others (Massey and Meegan 1982, 1985). Thus the context and the company encouraged inquiry into issues that were policy-oriented and socially relevant. Concern about societal problems was not enough; we were encouraged on a daily basis to take part in public debate. We felt comfortable in and received encouragement to pursue research projects on contemporary problems.

Confronting economics and economists

To understand how activism and public engagement came to be a more or less normal part of my cohort's experience, it is essential to understand the peculiar moment of the 1980s and the position of the field of economics in public policy in the United States (as distinct from the United Kingdom). Today when I go to policy meetings it is exceedingly rare to find a political scientist, or sociologist or planner, and it is even rarer to find a geographer.

Today, in the United States, neoclassically trained economists have a lock on policy discourse. This was not always the case. In the 1980s, neoclassical economics was not monolithic. In some circles that world-view was being called into question. In the 1980s, an era of unprecedented industrial restructuring, 'legitimate' critics from the left, center and right challenged status quo explanations for the nation's economic woes. In fact, it was an incredible moment when the likes of Bennett Harrison and Barry Bluestone, and less controversial figures such as Lester Thurow (Thurow 1981, 1984, 1985a, 1985b, 1993; Thurow and Tyson 1987), Robert Reich and Ira Magaziner, Wall Street investment banker Felix Rohaytn and political scientist Chalmers Johnson of University College Berkeley, coalesced around a set of arguments that raised the spectre of failure in the United States model of market capitalism (Johnson 1982, 1984, 1987; Johnson et al. 1989; Magaziner and Reich 1983). The Japanese and the emergent Asian Tigers were encroaching upon United States industries such as autos, computers, and clothing, and thumping national firms. Other models of capitalist development were not just curiosities, but instead were discussed as competing alternatives to the United States system of market capitalism. The failure of the United States system was increasingly being laid at the feet of United States corporatism.

A whole new debate unfolded about whether America should pursue industrial strategies to maintain its competitiveness. Berkeley professors and the Berkeley Roundtable on the International Economy, an influential university-based think tank, were important influences on and somewhat 'neoliberal' voices in the late 1980s. While more concerned about the social consequences of change, many students were funded through the Berkeley Round Table on the International Economy (BRIE) and represented in effect an institutionalization of what looking back must be now considered a pretty radical conversation (see, e.g. Johnson 1982, 1984, 1987; Johnson et al. 1989; Tyson 1992; Zysman and Tyson 1983).

With a Democratically-controlled Congress and a Federal agency apparatus populated by liberal social scientists, a moment of self-doubt and indecision descended upon the economic policy establishment. Questions were being raised about whether there was a better way to organize the economy and society. In this conceptualism there was a positive and active role for the state combined with the greater involvement of citizens and local organizations. Admittedly, the one weakness of the time was the failure to articulate a comprehensive and actionable alternative to 1950s Keynesianism. The diagnosis of the problem was only one of the steps required to mend the national economic condition.

This unique moment allowed a range of voices to be heard, among them geographers, planners and more institutionally minded economists. Becoming an academic during this time was easily coupled with a belief that a person could make a difference and could profitably contribute to policy discussions.

20 years later

Since the late 1980s, there has been a complete reversal in the political consciousness of the nation. It is no longer easy to find legitimate critiques of the status quo, particularly in policy circles in Washington. The economic crisis in Japan and the stagnant labor market in Germany diminished enthusiasm for and belief in the efficacy of alternative economic paradigms. In the early 1990s, Washington policy discourse became dominated by ideologically driven think tanks with tremendous sums of money deliberately deployed to shape policy conversations in the nation's capital.[2] Organizations like the Heritage Foundation and the Cato and American Enterprise Institutes all have budgets in the millions of dollars and dwarf liberal policy research think tanks like the Economic Policy Institute and Brookings, with large well-paid and well-supported research staff and savvy media consultants aggressively weighing in on contemporary issues in a sustained manner.

Progressives hoped that the Clinton administration would help reverse some of the regressive social policies promulgated in the Reagan era. But soon after Clinton entered the White House there was further erosion of liberal ideals as conservatives made inroads in a number of areas and neoclassical economic reasoning began to once again dominate policy discourse. The problems that confronted the Clinton administration on the eve of taking office and the increasingly important role played by Wall Street financial advisers in national politics and economic policy enabled Neoliberalism's creeping reach to define both macro economic and domestic policy designs.

The Clinton administration ushered in many important policy innovations in the areas of housing, environment, and labor policy, but by the end of the second year of the first Clinton administration, the die was cast. The remainder of the 1990s consisted of Democratic attempts to hold the line against Republican encroachment on liberal policy values and goals. By the late 1990s, economists once again ruled policy discourse, and there was little room for alternative conceptualizations of policy problems, let alone practice.

Where were the geographers during this time?

The economic collapse of the early 1980s unleashed a search for solutions to local economic development problems. President Reagan's inaction in the face of serious regional decline forced states and local governments to seek their own solutions to economic crises. The rise of high-technology industries, the much heralded 'death of the big firm', and the discovery of industrial districts (and their presumed behavioral underpinnings known as flexible specialization) emerged as interventions in local and state policy discourse. Entering the discussion later (compared with Europeans), United States economic geographers offered explanations for the problem of industrial transformation and in some cases were also consulted about solutions to job loss and industrial decline. During this time economic geographers provided some of the rhetoric that fueled policymakers' enthusiasm for things small, linked, clustered, and the like.

Ironically, geographers came to uncritically support these economic 'discoveries'. A substantial body of literature from the previous decade uncovered little relationship between industries that were co-located and had strong inter-industry linkages and vice versa (Chinitz 1960; Cooper 1971; Cromley and Leinbach 1980; Erickson 1972, 1973, 1974, 1975, 1976; Fagg 1980; Gordon 1987; Hagey and Malecki 1986; Hansen 1980; Haug 1981; Hoare 1985; Leone and Struyk 1976; Mulligan 1984; Oakey 1979a, 1979b; O'Farrell and O'Loughlin 1981; Struyk and James 1976; Thomas and LeHeron 1981).[3]

So enthusiastic was the adoption of districts, clusters and the like that when the edges of the argument about Marshallian districts began to fray and careful research demonstrated only loose associations among proximate firms, many economic geographers ignored such findings. With much at stake and policy audiences willing to listen to stories with happy endings geographers were surprisingly uncritical of the largely unsubstantiated body of research on districts and clusters. No doubt the places from which the original ideas emerged embodied the fabled characteristics of linked industries, but the empirical verification of the replicability of unique places was sorely lacking (Glasmeier 1987; Gordon 1987; Massey et al. 1992; Roberts 1972; Segal et al. 1985; Shapero 1972; Shapero et al. 1965). It would be several years before the peculiar non-economic factors were unearthed and made obvious. By then, policymakers had uncritically bought hook, line, and sinker the idea of clusters, linked industries, and industrial districts. Criticisms were ignored and evidence went unheeded. It would be almost ten years more before surveys and additional case studies offered enough evidence to suggest the fragile nature of the original hypothesis. Unfortunately, this compilation had not occurred before hundreds of communities, states and even national governments adopted programs designed to privilege certain industries. Belated commentaries on the likelihood of replicating unique place-based development experiences came too late.

Our own zeal returned to haunt us as policymakers and other advocates (Porter's Institute for Competitive Inner Cities (ICIC), The State of Arizona, The US Department of Commerce) ignored economic geographers' critiques and sought people who would tell them what they wanted to hear (for a critique see Garvin 1983; Fuellhart and Glasmeier 2003; Glasmeier 1999). We unwittingly became servants of a policy perspective that turned problems once described as regional misfortune into the practice of regional competition in which few places could hope to succeed. Economic geographers were listened to as long as they said what others wanted them to say. During this period, policymakers chose to ignore exhortations about probabilities and likelihoods. Those who criticized these overly optimistic tales of development were replaced by others who would reiterate what policymakers found palatable, if unattainable.

Debates about the efficacy of such policies did not lead to evaluation research that would have put muscle behind the critique. Clearly, without the evidence needed to support single topic strategies of development, economic geographers can never be serious policy analysts. When policymakers stop listening to our warnings, economic geographers should have turned to verifiable, critical analysis. Unfortunately, policy analysis skills are required to stay active in this type of debate. It is too often the case that economic geographers infrequently exercise evaluation skills that can be brought to bear on public debate.

Policy analysis: what is it, who does it, and how is it done?

Economic geographers do not engage in the policy process because they lack the skills to do so (see Staeheli and Mitchell 2005 for a discussion of geographers and policy participation). Policy research is about evidence and is based upon a specific methodology. If we are not prepared to challenge the beast on its own terms, with statistical models and conventional representations of the world, then our contributions will be limited to description. This is good as far as it goes, and in the initial stages of a policy trajectory it is a fundamental place to start. But, to really make a lasting difference, one that improves the lives of people and the health of communities, we must go beyond description and subject our initial hunches to rigorous evaluation, even at the risk of discovering that they are ultimately relevant only in very specific contexts. Rigorous analysis is the only way we can escape from being handmaidens to a policy process that is fraught with unequal power and poorly understood problems (see people like Jennifer Wolch whose research reflects a contemporary example of effective policy research). People like Bennett Harrison and Barry Bluestone used data and statistics to make their interventions. What set them apart now is their ability to listen to and critique policy dialogue on its own terms. Similarly, what set the activist-oriented geographers of the 1980s apart was their training in conventional theoretical approaches and subsequent decision to challenge them.

Economics is not the only issue

Economists are the dominant advisers in policy discussions today, in part because of their positive view of the world and in part because policy problems are increasingly narrowly scripted and framed in a manner that excludes questions that are not affirmative. Stated another way, policy discourse is about how to bring into alignment the world as it has been defined by a narrow band of interests.

The ascent of economists as hegemonic policy wonks still does not entirely explain the absence of geographers in contemporary policy debates. At least since the beginning of the twentieth century, with minor and temporally specific exceptions, geographers have been silent on important social issues. There is a singular absence of discussions in the geographic literature about such issues as the Great Depression, the First and Second World Wars, Vietnam, the War on Poverty, and even the 1960s urban crisis.

Two years ago I explored geographical perspectives on a critical social issue – poverty in America. I took the top eight journals in geography published electronically in JSTOR,[4] the online full text article service, and asked a simple question: how many times was the term 'poverty' mentioned in the tens of thousands of words in articles published over the 80 or so years for which key journals existed? After conducting a complete search of the eight journals referenced in JSTOR I then expanded my search to the top 20 geography journals. What I found was nothing less than shocking: Over 80 years of journal entries and thousands of pages of articles, there were 700 uses of the term 'poverty'. In the top-ranked journals over the same period only 200 references were found. Half of the time the term, 'the poverty of knowledge', was used as a literary device. I found far fewer references to the spatial location of and explanations for enduring poverty. Over the same time period literally thousands of references to the term poverty could be found in the sociology, political science and economics texts. In the 1960–1970s, arguably the most active and well-funded period of social policy research focusing on issues of poverty and deprivation in the last 40 years, entries about poverty in sociology and economics journals number in the thousands. Evidently, it is not just that economists and sociologists have carved out a role for themselves in this area, but that geographers have chosen not to study problems like poverty in society.

Tracking poverty discourse carefully from the 1950s forward, I could find a few notable geographers actively engaged in policy research and referenced in the field-defining journals. Names do come to mind: Dick Morrill of the University of Washington; Stan Brunn of Kentucky; Bill Bunge and his various institutional associations; Brian Berry, then of Chicago; and Richard Peet of Clark University. Of a more recent vintage, Jan Kodras, J. P. Jones, and a few others also come to mind. A clinical assessment of geographers' participation in policy discussions of poverty pull up names that include Niles Hansen (an economist), Andrew Isserman (a regional economist), and a few others.

Relevance versus glamour: important policy problems are not always the most attractive

How do we explain the scarcity of poverty references in the geography literature? There is probably no more geographical problem than the origins of and persistence of poverty. Dating back to the New Deal and accelerating in the 1960s, policymakers have defined the problem of persistent poverty from the perspective that people are poor for many reasons. One deemed most important is where individuals find themselves, that is, where in space they reside. Starting with debates in the 1950s, policymakers in Washington argued for and eventually formulated policies that led to programs designed to address the existence of poverty in particular places. The southern coal fields of Illinois, the mountains of Appalachia, the copper belt of Michigan, the old textile region of the Northeast, the Mississippi Delta, and the Border region, all have been subject to policy discussions since the late 1960s. Why has there been no geographical traction for this subject?

The answer may be found, again, in history. The 1960s was a unique moment in the history of the discipline. The creation of a number of federal research programs led to the completion of a large number of studies on regional differentiation. Economists did not dominate the scene then, so what happened? I cannot help but wonder whether our quest for legitimacy in the larger scholarly community, combined with our destructive internecine rivalries, simply absorbed our attention. It also may be that the failure of theory and empirical observation to fuse in a way that could be understood by policymakers and concerned citizens left us with half a loaf; good description but no explanation. The timing is right in the sense that the discipline was busy fighting within itself to assert a single unifying theory, which as we know was not possible then and remains impossible even today. But do we need one?

The era of active spatially specific policy formulation set into place instruments of change that are now embedded and which we can study and understand. From their beginning, these policy instruments used terms to describe underlying problems that include resource exhaustion, institutional balkanization, structural economic change, political exclusion, social isolation, and discontinuity. These terms are all found in geographical research and undergird many geographical explanations of contemporary reality. If we study these subjects, why is it so hard to apply them as explanations for important geographic outcomes?

If truth be told, although economic geographers might not have considered many of these issues, others in other areas of the discipline have: Charles Aiken of Tennessee on spatial inequality and institutional racism as seen in the structure of southern urban settlements post reconstruction; Joe Darden of Michigan State and the spatial implications of institutionalized racism, housing policy and land use; and others. Further, there are numerous examples of research on inequality in the Global South by critical human geographers (Michael Watts as an example). I come away from this wondering whether we have carved ourselves up to be so distinct that commonality cannot be found and sympathetic conversations cannot occur.

Conversations from within the field

The question being asked here or the situation for which I am seeking answers is not mine alone. Recent articles about the relevance of geography, the role of geographers, the opportunities for a geographic point of view, abound in the flagship journals (see Peck 1999). (It is easy to leap back in time and find similar articles lamenting the lack of relevance of the field in social policy circles. It is not that we are just talking about being relevant; indeed, there is plenty of conversation about this very issue and abundant examples of geography's relevance. I seek to ask and answer a different question: Why, when we speak up, are we not heard? I can offer my own experience and the experience of others whom I have observed.)

Once again a caveat is in order. My colleagues in the United Kingdom, Europe, and to some extent Asia and Africa seem well-placed to at least engage in policy discourse. Perhaps this is because problems of underdevelopment, exploitation, and unevenness are regularly part of the conversation. Perhaps it also reflects the reward structure in places where an academic's living wage is not assured (Europe, and southern Europe in particular). It might also be that these national governments take issues of inequality and uneven development more seriously than we do in America. Further, perhaps other national governments do not see such a rigid distinction between members of the academy and the policy community. A deeper history might also serve to reveal intellectuals' different cultural predispositions to contribute to discussions about daily life. All of these possibilities are not conventions in the United States where policy is done by people specifically trained in that realm. The ordinary citizen is not in fact expected to engage in the political process. That is what representative democracy is supposed to be about – select your representatives and they will act on your behalf. This model of engagement though seemingly representative reduces, if not outright takes away, the need to contribute to public discourse. Perhaps this helps explain in part why American academics often find it difficult, if not distasteful, to participate in public commentary.

All in all, geographers are not heard because they see the complexity in all problems. This ability to see all sides of an issue is our strength and greatest liability. It is strength because the world is complex and most other realms reduce life to generalizable levels of abstraction. In doing so, however, we lose the pattern and quality of differentiation that reflects the true reality of the world. Somehow, in our recognition of difference, we are typecast as actors unable to generalize and make sense of the patterns we see.

Summary

All that is discussed here can be quickly restated; academic engagement in the political process best occurs when society is gravitating in that direction. It is a simple idea predicated on the knowledge that actions occur most easily when met with the least resistance. There are moments when many streams come together in a confluence of ideas that result in an intellectual consensus about critical problems to which are profitably aimed best efforts and significant energy. The 1980s was such a time. Underlying all of geography is the belief at some level that each situation is somewhat or somehow unique and should not be smoothed or glossed over – our regional roots confine our comfort level to the known and the knowable even as the world increasingly demands that we speculate about what is happening and why.

The events of Hurricane Katrina in August 2005 have awakened a sleeping giant, the longstanding denial of human dignity to many Americans who, because of their geography, are subject to circumstances that can only be described as inhumane and deplorable. The last time this giant was awakened was 40 years ago; geographers sat largely silent and on the sidelines. There is no need to now. We should make our views known, let our voices be heard and use our intuitions and analytical sensibilities, tools and knowledge of history to make a difference. This is our second chance – we cannot let it pass unnoticed or unheeded.

We also have to stop following the next fantastic idea. 'Creativity' abounded in New Orleans as the most indelible fact shaping the public image of the place. 'Creativity' coexists with inequality and has done so throughout history (see Florida 2005 for a discussion of creativity). Let's not get dragged into yet another unsubstantiated discussion that can only serve to enliven regional competition that pits one place against another. Instead, let's speak about understanding and cooperation. Let's get in and get our hands dirty.

Notes

1. A debate was launched about the failure of economic geographers to engage policy by Jamie Peck in 1999 in *Transactions* (Peck 1999, 2000; Pollard et al. 2000).
2. Jamie Peck and Adam Tickell's research on Neoliberalism highlight this development.
3. Richard Gordon, Political Economist at the University of California Santa Cruz, was an early critic of unbridled enthusiasm about the prospect of recreating Silicon Valley. He and his wife, Linda Gordon, undertook some of the early survey-based research of Silicon Valley supplier firms in which they demonstrated that even local firms were only marginally attached to one another and to the complex itself. Gordon's work showed the early international quality of the Valley and its supplier system.
4. An electronic, full-text article retrieval service available in many public universities.

References

Chinitz, B. (1960) 'Contrasts in agglomeration: New York and Pittsburg', *American Economics Association*, 51: 279–89.

Cooper, A. (1971) 'Spinoffs and technical entrepreneurship', *IEEE Transactions on Engineering Management*, EM, 18(1): 2–6.

Cromley, R. and L. Thomas (1980) 'The pattern of the filter down process in non-metropolitan Kentucky', *Economic Geography*, 57(3): 208–24.

Erickson, R. (1972) 'The lead firm concept and regional economic growth: an analysis of Boeing expansion, 1963–1968', *Tidjschrift Voor Economie En Sociologie Geografie*, November/December, 426–37.

Erickson, R. (1973) 'The lead firm concept and regional economic growth: an analysis of Boeing expansion, 1963–1968', unpublished dissertation, Department of Geography, University of Washington, Seattle.

Erickson, R. (1974) 'Regional impact of growth firms: the case of Boeing, 1963–1968', *Land Economics*, 50: 127–36.

Erickson, R. (1975) 'The spatial pattern of income generation of lead firm, growth area linkage systems', *Economic Geography*, 51(1): 17–26.

Erickson, R. (1976) 'The filtering down process: industrial location in a non-metropolitan area', *The Professional Geographer*, 28(3): 254–60.

Fagg, J. (1980) 'A re-examination of the incubator hypothesis: a case study of greater Leicester', *Urban Studies*, 17: 35–44.

Florida, R. (2005) *The Flight of the Creative Class: The New Global Competition for Talent*. New York: Harper Business.

Fuellhart, K. and A. Glasmeier (2003) 'Acquisition, assessment and use of business information by small- and medium-sized business: a demand perspective', *Entrepreneurship and Regional Development*, 15(3): 229–52.

Garvin, D. (1983) 'Spinoffs and the new firm formation process', *California Management Review*, 25: 3–20.

Glasmeier, A. (1987) 'Factors governing the development of high tech industry agglomerations: a tale of three cities', *Regional Studies*, 22(4): 287–301.

Glasmeier, A. (1999) 'Territory-based regional development policy and planning in a learning economy: the case of "real service centers" in industrial districts', *European Urban and Regional Studies*, 6(1): 73–84.

Gordon, R. (1987) 'Growth and the relations of production in high technology industry', *The Future of Silicon Valley*. London: George Allen and Unwin.

Hagey, M. and E. Malecki (1986) 'Linkages in high tech industry: a Florida case study', *Environment and Planning A*, 18: 1477–98.

Hansen, N. (1980) 'Dualism, capital-labor ratios and the regions of the US: a comment', *Journal of Regional Science*, 20: 401–3.

Harrison, B. (1997) *Lean and Mean: The Changing Landscape of Corporate Power in the Age of Flexibility*. New York: Guilford Press.

Harrison, B. and B. Bluestone (1988) *The Great U-Turn: Corporate Restructuring and the Polarizing of America*. New York: Basic Books.

Harrison, B. and A. Glasmeier (1997) 'Why business alone won't redevelop the inner city: a friendly critique of Michael Porter's approach to urban revitalization', *Economic Development Quarterly*, 11(1): 28–38.

Harrison, B., L. Baker and B. Bluestone (1980) *Corporate Flight: The Causes and Consequences of Economic Dislocation*. Washington, DC: Progressive Alliance Books.

Harrison, B., L. Baker and B. Bluestone (1982) *The Deindustrialization of America: Plant Closings, Community Abandonment, and the Dismantling of Basic Industry*. New York: Basic Books.

Haug, P. (1981) 'US high technology multinational and Silicon Glen', *Regional Studies*, 20: 103–16.

Hoare, A. (1985) 'Industrial linkage studies', in M. Pacione (ed.) *Progress in Industrial Geography*, pp. 40–80. London: Croom Helm.

Johnson, C. A. (1982) *MITI and the Japanese Miracle: The Growth of Industrial Policy, 1925–1975*. Stanford, CA: Stanford University Press.

Johnson, C. A. (1984) *The Industrial Policy Debate*. San Francisco, CA: ICS Press.

Johnson, C. A. (1987) *The Political Economy of the New Asian Industrialism*. Ithaca, NY: Cornell University Press.

Johnson, C. A., L. D. Tyson and J. Zysman (1989) *Politics and Productivity: The Real Story of Why Japan Works*. Cambridge, MA: Ballinger.

Leone, R. and R. Struyk (1976) 'The incubator hypothesis: evidence from five SMSAs', *Urban Studies*, 13: 235–331.

Magaziner, I. C. and R. Reich (1983) *Minding America's Business: The Decline and Rise of the American Economy*. New York: Vintage Books.

Massey, D. B. and R. Meegan (1982) *The Anatomy of Job Loss: The How, Why, and Where of Employment Decline*. London: Methuen.

Massey, D. B. and R. Meegan (1985) *Politics and Method: Contrasting Studies in Industrial Location*. New York: Methuen.

Massey, D. B., P. Quintas and D. Wield (1992) *High-Tech Fantasies: Science Parks in Society, Science, and Space*. New York: Routledge.

Mulligan, G. (1984) 'Agglomeration and central place theory: a review of the literature. *International Regional Science Review*, 9(1): 1–42.

Oakey, R. (1979a) 'The effects of technical contacts with the local research establishments on the location of the British instruments industry', *Area*, 146–50.

Oakey, R. (1979b) 'Labor and the location of mobile industry: observations from the instruments', *Environment and Planning A*, 11: 1231–40.

O'Farrell, P. and J. O'Loughlin (1981) 'New industry input linkages in Ireland: an economic analysis', *Environmental and Planning A*, 13: 285–308.

Peck, J. (1999) 'Grey geography?' *Transactions of the Institute of British Geographers*, 24: 131–5.

Peck, J. (2000) 'Jumping in, joining up, getting on', *Transactions of the Institute of British Geographers*, 25: 255–8.

Pollard, J., N. Henry, J. Bryson and P. Daniels (2000) 'Shades of grey? Geographers and policy', *Transactions of the Institute of British Geographers*, 25: 243–8.

Roberts, E. (1972) 'Influences on the performance of new technical enterprises', in A. Cooper and J. Komives (eds) *Technical Entrepreneurship: a Symposium*, pp. 126–49. Center for Venture Management., Milwaukee, WI.

Segal, Quince, and Partners (1985) *The Cambridge Phenomenon: The Growth of High Technology Industry in a University Town*. Cambridge, UK: Segal, Quince, and Partners.

Shapero, A. (1972) 'The process of technical formation in a local area', in A. Woper and J. Komives (eds) *Technical Entrepreneurship: A Symposium*, pp. 63–96. Center for Venture Management, Milwaukee WI.

Shapero, A., R. Howell, and J. Tombaugh (1965) *The Structure and Dynamics of the Defense R&D Industry: The Los Angeles and Boston Complexes*. Menlo Park, CA: Stanford Research Institute.

Staeheli, L. and D. Mitchell (2005) 'The complex politics of relevance in geography', *Annals of the Association of American Geographers*, 95(2): 357–72.

Struyk, R. and F. James (1976) *Intrametropolitan Industrial Location*. Washington, DC: National Bureau of Economic Research.

Thomas, M. and R. LeHeron (1981) 'Perspectives on technological change and the process of diffusion in the manufacturing sector', *Economic Geography*, 51(3): 231–51.

Thurow, L. C. (1981) *The Zero-Sum Society: Distribution and the Possibilities for Economic Change*. New York: Penguin Books.

Thurow, L. C. (1984) *Dangerous Currents: The State of Economics*. New York: Vintage Books.

Thurow, L. C. (1985a) *The Management Challenge: Japanese Views*. Cambridge, MA: MIT Press.

Thurow, L. C. (1985b) *The Zero-Sum Solution: Building A World-Class American Economy*. New York: Simon and Schuster.

Thurow, L. C. (1993) *Head to Head: The Coming Economic Battle Among Japan, Europe, and America*. New York: Warner Books.

Thurow, L. C. and L. Tyson (1987) *Adjusting the U.S. Trade Imbalance: A Black Hole in the World Economy*. Berkeley, CA: Berkeley Roundtable on the International Economy. Available at: http://repositories.cdlib.org/brie/BRIEWP24, accessed 22 January 2006.

Tyson, L. D. (1992) *Who's Bashing Whom?: Trade Conflicts in High-Technology*. Washington, DC: Institute for International Economics.

Zysman, J. and L. Tyson (1983) *American Industry in International Competition: Government Policies and Corporate Strategies*. Ithaca: Cornell University Press.

18 The new imperial geography

John Lovering[1]

A geographical–economic question? Not sure we can help

Imagine you are a community representative, a businessperson, or just curious, and you want to ask questions like: 'How does this place (region, city, small country) work? Why can my daughter get a job here when my son can't? What can be done to make things better?' You might think the best person to ask is an economic geographer.

Economic geographers tend to fall into two distinct camps. Members of the first would typically respond with a species of dazzling poetry about how fascinating diversity is, how everything is all mixed up, how it looks different depending on who you are, and how there is no last word (as if you didn't know that already). Members of the second would scrub out your questions, replace them with one about 'competitiveness', then answer it by declaring that public resources should be diverted to give special help to this or that set of special interests.

Of course, most economic geographers are decent folk and wouldn't do either of these so crudely. But many would feel it professionally prudent to make at least a nod towards one or both. For these two pole positions in Post-Cultural-Turn Economic Geography (henceforth PCTEG) preoccupy the attention of publishers, university appointment committees, and funding bodies. Yet neither constitutes progress in any familiar sense of the word because they do not answer questions any better than in the past. They are about asking different questions altogether. This is not, despite pop interpretations of Kuhn, how sciences get better. For example, in their recent survey Barnes et al. (2004) note that the story of recent change in economic geography is not one that everyone agrees signifies progress, and ask how we should interpret it. This chapter offers one interpretation: that it reveals geography's excessive embrace of the Empire of Capital.

This has nothing to do with the wonderfully widened range of topics (there's nothing *inherently* Imperialist about studying gardening). The complicity arises from the cognitive and normative frameworks within which these are all too often set, which smuggle in Empire as the un-named, unconscious, horizon of authorised thought and practice. Since this is an Empire characterised by denial, this is achieved through ideas presented as inherently anti-foundational, critical, destabilising, engaged, inclusive, and other labels giving the impression that they are definitely *not* part of a conservative orthodoxy, like rebel clothing in designer shops.

The new imperialism: the proliferation of networks, difference, states and markets

Humankind has lived most of its recorded history under Empires. The novel features of the latest version[2] are most lucidly set out by Ellen Meiksins Wood (2003). Most novel of all, it doesn't officially exist. This is an Empire 'administered by a global system of multiple states and local sovereignties' (Wood 2003: 141). It colonises by annexing not territory but the thinking and behaviour of a multiplicity of policymakers at a variety of scales. So there are no formal imperial institutions, merely a shifting constellation of corporations, border-crossing networks, and territorially-defined political units representing, or at least ruling over, distinct 'communities'. A little local colour, a plurality of perceived identities and of governments are more than curiosities, they are essential. The global convergence of policy thinking and outcomes draws on, and fuels, the construction and mobilisation of difference.

Rather than extracting resources from formally subject peoples, Imperial prosperity is derived from the extension of the arena of capitalist accumulation. This is being further extended through the spread of neo-liberalism,[3] the Hayekian project to socially-engineer societies and individuals to make them fit market forces (Saad-Filho and Johnston 2005). Under this influence the 'global labour supply' accessible to mobile capital increased by about a third in the 1990s. Neo-liberalism is compatible with enormous diversity, attaching itself to religious fundamentalist politics here, and populist consumerism there. But its encroachment is associated with an easily recognisable common core of mechanisms and manifestations (from the commodification of health and pensions to 'active labour market policies'; from instrumentalist education to strategies to manage the 'socially excluded'; from place marketing to shopping malls with the same shops, from a spate of new tall downtown buildings to Starbucks, and so on).

As neo-liberalism spread, global per capita economic growth rates declined (now at a third of the 1960s), North–South resource flows went into reverse, and in many places so did equalising tendencies between the genders. Reduced growth, increased imports and inward investment and consequent market saturation intensified the struggle for market shares. This in turn triggered a corporate (and thence governmental) obsession with innovation, and a huge expansion of advertising and marketing expenditures. The diversion of investment towards finance markets and property is now fuelling a spiralling of personal debt (in the 'West') and a spectacular new round of urban 'regeneration', currently transforming the visual and social character of the world's cities (Smith 1996). The American model of the city, not one of urbanism's greatest successes, is being copied everywhere.

A commodity cornucopia coupled to market differentiation exploiting the commodification of difference has created a new fusion of identity, consumption and lifestyle. But the shift towards a more 'vertical' pattern of accumulation has also produced unprecedented levels of inequality and poverty, and a globalisation of 'Western' afflictions such as depression and urban fear. Some tentative evidence suggests that beyond a low threshold ($15,000 per annum) increases in income bring severely diminishing marginal gains in happiness. In the rich countries economic welfare seems to have been declining ever since neo-liberalism arrived (Layard 2005). Some rather more detectable evidence points to a possible environmental catastrophe. The Empire is a frenetically busy and glitzy place, but not a fair, happy or sustainable one.

Since neo-liberalism requires, contrary to its sales rhetoric, 'extensive and invasive interventions in every area of social life' (Saad-Filho and Johnston 2005: 4) its key element is the nation-state. Its numbers have quadrupled since 1947, much of this during the neo-liberal period. The number of sub-national units of governance has multiplied even more, enabling the recruitment of locally defined identities to strategies for 'competitiveness'. Along with the on minimal taxation on the well-off, this has levered tens of thousands of private companies and non-governmental organisations into the marketised business of governance. A crucial aspect of Empire accordingly being the construction of ever more 'networks'.[3]

The American flavour of the Empire of Capital has given rise (especially since the invasion of Iraq) to racist anti-Americanism and much finger wagging at the Bush administration. But its roots lie in the gradual consolidation through the twentieth century of external and internal conditions whereby the uniquely gigantic US state became able and willing to play the role of Hobbesian planetary Leviathan. Since the 1970s the United States has been the first major victim of, then the main exporter of, this particularly voracious form of capitalism (Harvey 2004). The United States now plays the leading role in promoting, through regime changes both formal and informal, a world of 'market states' (Bobbitt 2002; Ferguson 2004).

In this perspective, the paradigmatic new Imperial event is not the unleashing of high-tech military violence by Americans thousands of miles away from their victims. It is one of those routine conferences in which political leaders, academics, consultants, and business people make speeches about how to transform this or that real or imagined aspect of their locality into a marketable asset in the struggle for 'competitiveness'.

An economic geography for the new empire

Every Empire is served by a stratum of intellectuals who have specialised in manufacturing consent, and this one is no exception. At the 'high-theory' pole this consists of a theological elite – a Priesthood – who articulate and legitimate the broad cognitive and normative assumptions that render Empire as destiny, or at least, unavoidable. At the humbler, practical, pole it consists of a mass of policy 'artisans' who play an important role licensing and lubricating the Imperial project in specific local situations, helping to construct the specific channels and projects through which hegemonic perceptions and imperatives find their way into governance at all scales.

The academic ideas of the age tend to be closely related to those of the most powerful groups. So it is not surprising that 'Post Cultural Turn Geography' has generously partaken of, and sometimes added a few tweaks to, these classic Imperial practices. In parallel with the global projection of neo-liberalism and rise to unchallenged hegemony of the United States, a combination of postmodernism and post-structuralism became the cuckoo in the official intellectual nest (Callinicos 1999: 297). Anglo-American human geography was one of the most thorough and lasting conquests (Soja 1989). The post-modern element now looks more than a shade old fashioned, neo-modernist grand narratives being louder and more monolithic than ever (notably that there are no alternatives to neo-liberal economics and US-style 'liberal democracy'). But its post-structuralist cognitive and normative foundations live on the prevailing dominant academic (and non-academic) orthodoxy in the West (and increasingly elsewhere). They have licensed an evolving sequence of post-post-modern discourses from Actor Network Theory, Relationality, to non-representational 'theory', and beyond.

Geography's animal farm

In the 1970s many of the up-coming generation of geographers, especially in the United States and the United Kingdom, felt there was something profoundly rotten in the discipline. To cut a long story very short, they regarded it as excessively complicit with big business, the state, environmental plunderers, or more fundamentally patriarchy and capitalism. And they traced this to its 'positivist' philosophical underpinnings (though 'positivist' was never quite the right word). The critique held that by reifying 'empirical' categories this in effect limited the geographically thinkable to the concerns of dominant interests and ideologies. If the apparently-obvious is all there is, any attempts to explain by reference to less obvious forces (capitalism, patriarchy, ideology, Empire, etc.) are ruled out as pseudo-scientific or quasi-mystical. Over the following two decades what many claimed to be a Kuhnian paradigm change opened up wider research agendas and techniques (and those concepts became respectable and researchable). But by the 1990s and 2000s, when my generation was settling in at geography's commanding heights, while neo-liberalism was rampant both outside and within the academy, the new had become the old, the revolution had faded into an orthodoxy, and geography was returning to business as usual – the delights of fetishising place and space.

And this was once again rationalised by empiricism, albeit a re-vamped version. This time empiricism drew on a depthless ontology of infinite points and lines, networks, or performativities, waving a license signed by Deleuse and Derrida (possibly a forgery). The new epistemological orthodoxy ironically disinterred an idea from the most degenerate and dictatorial post-Classical phase Marxism: that

Truth is the discourse that favours the working class (but minus the working class). 'All science is ideological, only we admit it, and we will not let the facts get in the way of our favoured stories' (Sayer 2000: 59). This ruled out the possibility of thinking space as hiding and reproducing capitalist or Imperial power even more presumptuously than its white-coated number-crunching positivist predecessor.

The Priesthood: no thinking about a world beyond Empire

A classic statement of the highly foundational 'anti-foundationalism' characterising the PCTEG asserts:

> A capitalist firm cannot appear as the concrete embodiment of an abstract capitalist essence. It has no invariant 'inside' but is constituted by its continually changing and contradictory 'outsides'.
>
> (Gibson-Graham 1996: 15)

This speaks not of evidence or logic, but desire and cultural authority. Even if one buys (to use market-speak) into the first sentence as rejecting one kind of simple reductionism, the second is a non-sequitur. To insist that a firm can only be conceived of as a thing made up of 'outsides' is only to demand that the kinds of questions asked, and the kinds of answers research comes up with, be compatible with extreme ontological actualism (the empirical, the actual, and the real cannot be distinguished). Insisting that things really are only what they appear to be (if you squint this way or that) is of course philosophising of the most foundational, and authoritarian, kind.

Under the influence of this kind of thinking ontology and epistemology in much of economic geography has collapsed into a matter of choosing which Deleuzian plane of immanence you fancy you can skateboard to the horizon on. Geographical research accordingly becomes all about you, just like shopping.[4] Since this rules out the earlier scientific ambition of trying to find anything that might show some arrangements ought to be changed (as opposed to starting off with that prejudice in mind), research has to be justified by criteria other than discovery or claim-testing. Given the popularity amongst those doing the authorising of the notion that knowledges are entirely discursive constructions, and that the merits of different knowledges should be judged by the subjects, practices, and identities they 'empower', the answer was to see research as a matter of benign compilation. So the disciplinary corridors began to fill once again with (this time virtual) cabinets containing collections of specimens deemed, by their collectors, to be particularly worthy.

Where the geographical authorities of nineteenth century imperialisms gave names to colonised peoples and places, their equivalents in the twenty-first Empire of difference and markets invent names for imaginary spaces (folded, Third, Alternative, Resistant etc.), as if to fulfil that old prediction of the first time as tragedy, the second as farce.

Intoxicating textualism

Post-structuralism, at least as it has been most influentially imported into economic geography, elaborates Nietzsche's insistence that since every statement has a textual character, it's a good idea to ask who is speaking. Reworked in less authorial terms as a fixation on the problems of relating signifiers to signifieds, and signifieds to anything non-textual, this not terribly radical century-old observation has turned out to be extraordinarily intoxicating for geographers in the neo-liberal age. Perhaps this is because geography is a synthetic discipline in which theory-surfing and neo-orientalism are pathological tendencies.

The results vividly demonstrate the contradictions created by letting textual strategies substitute for, rather than be a part of, investigating the world. Anthony Easthope (2002: 4) argues that it's just not possible: 'for human beings, as speaking subjects, to encounter ... a gap in signification without immediately trying to close it' and they usually do so under the influence of subconscious desire. Economic geography certainly fits the description; its closures have typically reflected the habitus of the 1960s generation now in authority, in the form of feminist, environmentalist, animal liberationist, anti-racist (etc.) conventions. In the name of respecting difference these have installed one or other simplistic category of sameness (an identity).

The more theoretically sophisticated, or perhaps less politically-driven, thinkers have tried to avoid such contradictions by forever running ahead of closure, chasing and abandoning this then that discourse in an attempt to avoid any vocabulary of representation at all. But deferring meaning till the seminar is over does not overcome the problem that unless geography is to say nothing, some statements have to be made. And whichever the set of discursive conventions you opt for, some statements will – for all their contingency – be more practically adequate than others. This is not because, a la Rorty, the speaker – or 'the discourse' – is on the side you fancy. But because, under the description adopted, some will indeed 'represent' things – not perfectly – but better than others (Sayer 2000). Economic geography's textualist obsession has meant that this unremarkable notion, good enough for other sciences to get on with and produce ideas and technologies that work, has it spinning round in circles. It is still said in many a geographical lecture room that to make any Truth claims is to pretend to adopt an external Archimedian point of God-like objectivity – and this from geographers who write numerous articles and books describing how things are, i.e. make truth claims.

The 1960s originals were less inclined to lose the tune than their present day tribute acts. Foucault or R. D. Laing, for example never assumed that their passions for druggy peak experiences had any great significance for their analytical and political work. But the authorities of geography leading the empiricist, relativist, revival today are more smart-casually dressed, are much more extreme, endorsing a hazy blurring of the styles of myth, literature and science. This licenses nice storms in academic–geographical teacups. Meanwhile, outside the window, the property developers get on with building luxury flats all over the playing field.

If the Empire had a core of scheming magicians whose business it was to cook up ways of thinking that would render alternatives, and critical political engagement, unthinkable, they could hardly have done better.

Economic geography minus the economics

Thanks to these influences the PCTEG is economic only in a thematic sense (it uses the word 'economy' occasionally). It's students are unlikely to have any but a very tentative grasp of how economic theories are constructed and used, or of geographic–economic history beyond the comic book categories of Fordism, PostFordism, Knowledge Economy etc. They are unlikely to be familiar with the nuances of the debates over capitalism's macroeconomic and growth tendencies and its cyclical trends. Despite all the deconstructable ambivalences that run through those debates, like any, this means they are missing something.

Economics, from Adam Smith onwards, was distinguished by its concern with the emergent properties arising from organising production and consumption through markets. The common PCTEG claim that this meant treating the economy as a machine misses the point entirely. Economic inquiry (liberal or Marxist) aimed to explore *how* machine-like properties could arise; under what conditions can the whole turn out to be much greater than, and so different from, its parts? The classical economic tradition also gave a central place, at least in principle, to the cultural embeddedness of the economy, as even a superficial glance at Smith, Marx, Bukharin, Keynes, or Sen would show. For 'culture' defined what counts as a commodity and a market in the first place. You would never know this from the more extreme celebrants of the PCTEG, who present a neo-Whiggish graph of intellectual history in which it appears as the crowning peak, rather than another haze in a dip.

The PCTEG pretends it has turned economics' thematic questions and analytical insights to mush in the acid bath of textualism, and invented its cultural ones anew. But it's a con trick. When Thrift and Olds announce 'the full complexity of modern economies only become apparent when we move outside of what are often still, considered to be the "normal" territories of economic inquiry' (Thrift and Olds 2004: 59) they are not reporting a fact, merely demanding that attention be redirected from one kind of complexity to another that they prefer (often those where there is no chance that research might ever arrive at some kind of practically adequate answer). This doesn't necessarily bring complexity to view at all, though it does bring into view the prejudices of the authors. Ten years ago Stuart Hall warned of the danger of flipping from a naive economism to an even more naive culturalism.[5] PCTEG all too often proves his point.

Extracting from economic geography of any way of grasping arguments for this or that market intervention other than by a discursive reduction to those for or against, has left it with little grasp of economic development, or much to say about policy. Assertions by figures such as Castells, Sassen, and others, for example that a city or regions' past and future, and its policy priorities, are explained by its 'competitiveness', have been accepted as gospel. But they are theoretically incoherent (prosperity is a function of productivity, whether this corresponds to 'competitiveness' is entirely contingent on circumstances) and empirically question-begging (the development of the finance centre of London, for example, is unlikely to explain more than a tiny part of the London labour market, though it has affected its planning and its skyline). To take another example, the current global house price boom is a major factor in the current frenzied round of urbanisation and 'regeneration'. But the PCTEG offers neither the techniques nor the motivation for a macro analysis of circuits of capital (which might explore how this is connected to neo-liberalism's redirecting of government intervention from tax and job-creation to credit-driven consumption management), or a more local inquiry along the lines of the questions in the introduction (which might show how it is instantiated locally, and how it might be moderated a little here or there).

From empirics to empiricism

The loss of empirical substance (as opposed to a celebration of empiric*ism)* is the distinctive feature of the PCTEG. It was once 'normal science' in regional or urban economic research, or development studies (under the heading of an examination of the export base or the investment multiplier) to conduct empirical research to find whether this or that economic factor, actor, sector, or whatever had some special significance as a driver of economic change in a chosen place (and therefore perhaps of special policy interest) (Dicken and Lloyd 1978). The PCTEG has no capacity to perform on this stage, so it is often left to neo-liberal simplistes (see Artisans below) who in effect know all the answers a priori ('more competitiveness'!). An economic understanding of space and place at more than a sloganistic level requires the type of empirical (not *empiricist*) investigation for which the PCTEG provides neither skills nor respect. I take it this was one of Ann Markusen's (1999) points in the 'fuzzy concept' paper, much misrepresented as an old-fashioned call for quantitative approaches (Peck 2003).

Lancaster's famous (amongst economists) theory of the second best, showed that an intervention that would produce certain results in one context could lead to dramatically different ones if the starting conditions were only very slightly different. This would suggest that policies for globalisation, 'competitiveness', the 'knowledge economy' or some other claimed imperative need to be examined a lot more closely in specific context before anyone can have any confidence that they will deliver what they promise even in their own terms. This is demonstrated independent of anyone's preferred values. This kind of contribution is not possible within the PCTEG.

... and the geography

The second casualty of the great purge has been a similarly radical redirection of economic geography's geographical imagination, severing it from what most people think the subject is about – an empirically informed awareness of the planet we live on. The PCTEG monastery is a delightful place for unspecified timeless ponderings about liquidities, propinquities, and relationalities untrammelled by any pressure to weigh these on some kind of scales balanced by disprovable facts, or to work out rigorous theories. The results are segued, according to preference, to equally theological policy utterances, conjuring up agonistic engagements, fleeting coalitions, passions, empowerment, and other wonderful things that are unlikely to require any political body to actually change anything, except perhaps in the rhetoric of its procedures. To judge by the practice, the job of high-level economic geographical theorists today is to contemplate and pontificate, primarily amongst themselves. The notion that geography might seek to discover (rather than invent) and pass on some information that could be considered as sufficiently factual to change some minds, is history.

If persuasion is not to be achieved by presenting logic or evidence, cultural capital must do the job. Ash Amin (2004), for example recently asserted that globalisation and the general rise of a society of transnational flows and networks 'no longer allow' a conception of place politics in terms of spatially bound processes and institutions. This certainly illustrates the labour saving benefits of this kind of geography: just to mention relationality is enough to solve even the trickiest of problems, like the significance of scale, in a trice. There's no need to actually investigate anything anywhere. The new emphasis on relationality is motivated, according to Doreen Massey (2004), by the political need to combat localist or nationalist claims to place-based essentialisms. She does not explain whose need this is, nor why no equivalent campaign was needed to combat globalist essentialisms. Equally puzzling, this kind talk is not new. As Henry Yeung (2005) notes, geography has been 'relational' all along: that's what made it geography in the first place. The novelty today seems to lie in the implication that invoking it is the end, rather than the beginning, of geographical inquiry.

The PCTEG repeatedly demonstrates this irrational leap from the (elementary) observation that all empirical studies are theoretically conditioned to a cavalier abandonment of any careful empirical input at all. Amin and Thrift (2005: 238) for example recently delivered the judgement that geography is 'moving on', but unfortunately they gave no criteria whereby anyone other than themselves could distinguish between 'on', 'back', 'down', 'off' or 'nowhere'. They then issue a call to arms insisting on 'not only imagining the world in multiple ways but also a willingness to engage with heterodox thinking from all manner of disciplines'. But since they gave no clues as to what 'engagement' might mean, *how* to engage, on what *basis* such encounters should take place, only they will be able to judge whether these noble tasks have been accomplished. The model of geographical research presented here is to collect lots of pictures of the other in a gigantic photo album, the beauty of which only an expert can judge.

So it is no wonder that the cognitive light is so hazy in many geographical corridors that many find themselves embracing neo-liberalism without recognising it, or get to see only its most seductive profile. So the geographical friends of Empire now include a deconstructionist chapter. Doel and Hubbard (2002: 365) for example, dismissing the idea that a city might be a place, conclude that 'cities can only enhance their competitiveness by recognising that world cityness ... needs to be performed and worked at in a multiplicity of sites'. After scrambling through the textualist bushes they find themselves on the usual neo-liberal highway, trotting along with everyone else seeking 'competitiveness'.

The Artisans: no alternative to getting on with (imperial) business

Priestly deliberations provide the neo-liberal empire with the theological no entry signs to prevent sustained errant thoughts. But as noted earlier the Empire needs more focused interventions to convert its key mobilising concept, 'competitiveness' from a floating signifier to a set of practical policy guidelines in a particular situation. So down in the street tens of thousands of Artisans shape it's practical local implications and consequences (which industries, groups, and places, are to be favoured by what use of public resources, and how, etc.).

In complete contrast to the Priests, the Artisans (consultants, advisors, academics, journalists) are many in number but few in ideas, and can be identified by their membership of networks sharing a bulldozer-simple ontology: There Is No Alternative – to Competitiveness. The parables used to spread this – the demise of 'Fordism' or arrival of the 'Informational Society', 'Knowledge Economy', 'network paradigm' etc. – have never been developed with any rigour, and the shiniest are untarnished by anything vaguely approaching a historical fact (Henwood 2003). The idea that competitiveness is the only proper goal of economic development strategies, at whatever spatial scale, was borrowed from Michael Porter's US studies of the late 1980s (which were about firms) and misapplied to places. These stories make little economic or geographical sense at all. But they work well as devices whereby the cultural capital of academia (or think-tank) can be drawn down to impart gravitas to interventions in the policymaking arena.

The ideological veins and arteries pumping neo-liberalism around the Empire are most visible in these Porterist networks and discourses. Here the neo-liberal myth that entitlements depend on the ability to win a contest is spatialised, rationalising the privileging of one or other group of economic activities and actors. The issues have been extensively discussed under the heading of the debate over the 'New Regionalism' (Lovering 1999; Keating 2005) so there is no need to go over them here, beyond noting that the latest packaging is the slogan that 'cultural industries' are the key to 'regeneration'. It's impossible to put enough scare quotes in a sentence containing these terms.

Conclusions: what geography matters for

Geographers have long insisted that 'geography matters', although they have been conspicuously unsuccessful at pinning down exactly *how* (e.g. Soja 1989). One way to find out is to look at what geographers can actually do. Those trained in the arts of post-cultural-turn economic geography are conspicuously poorly endowed with the skills to do much that non-insiders (to the discipline or the Empire) are likely to regard as self-evidently useful, like answer the questions with which this chapter began. They are, however, marvellously qualified to talk the talk of Priests or Artisans. The most influential reconstructions of economic geography have turned it at one pole into a space for aesthetic-theological contemplation, expressing wonderment at all things spatial and different, and at another into a service industry mass-producing policy licenses and credentials. The former inclines to excusing or obscuring Empire, the latter to training its functionaries. Economic geography has made itself matter as an academic corollary and component of the Empire of capital.

That would not be a criticism if you don't mind about inequality or the longer-term effects of the neo-liberal global extension of capitalism, don't believe there's an 'us' worth talking about other than your preferred group, don't think anything can be done at any spatial scale other than to keep fingers crossed, or believe that the Empire will eventually turn out for the best.[6] But if you think economic geography should do more than play out orthodoxies, and should appeal to evidence and intellectual coherence as resources for transcending prejudice, then you will probably be looking forward eagerly to its next reinvention when another new generation arrives before too long.

Notes

1. Biography: my work life has consisted of spells in office work in nationalised industry, community development, rock music, and academia (economics, urban studies, geography, planning). My current transformation into grumpy old man is related to regular confirmation of the fact that while the capitalist the music industry must produce some popularly consumable use values if it is to realise any profits, the neo-liberal academic ideology and credential factory need not.
2. The *New* Imperialism is the subject of major debates but with the prominent exception of Harvey (2003) these have been from well beyond economic geography. Hardt and Negri (2001) drawing largely on the same inspirations as the PCTEG, offer an account which colours the Empire attractively green and red.
3. Networks became a fashionable topic in geography in the 1990s. The more sophisticated versions drew on Actor Network Theory (following Latour), which redefined the word 'act' to shed its usual connotation of intentionality (Fine 2002). Symmetry, 'actants' and power then turn up everywhere, but without any clear significance. The fashion for networks rendered Empire unthinkable just as it was being most energetically built – through the construction of networks.
4. The other post-structuralist, but ontologically deeper, tradition in Foucault, Lacan and Derrida that suggests that the main problem is taking this 'you' for granted, is out of tune with this neo-liberal-friendly version, and has received much less attention in the PCTEG, though see Massey 2004.
5. 'As if, since the economic . . . does not as it was once supposed to do, determine . . . in the last instance, it does not exist at all! (Hall 1996: 258)
6. Some combination of which seems to be the politics of the PCTEG.

References

Amin, A. (2004) 'Regions unbound: towards a new politics of place', *Geografiska Annaler*, 86(B): 33–44.

Amin, A. and N. Thrift (2005) 'What's left? just the future', *Antipode*, 37(2): 220–38.

Barnes, T., J. Peck, E. Sheppard and A. Tickell (eds) (2004) 'Introduction', *Reading Economic Geography*. Oxford: Blackwell.

Bobbitt, P. (2002) *The Shield of Achilles: War, Peace and the Course of History*. London: Penguin Books.

Callinicos, A. (1999) *Social Theory: A Historical Approach*. Cambridge: Polity Press.

Dicken, P. and P. Lloyd (1978) *Location in Space*. New York: Harper & Row.

Doel, M. and P. Hubbard (2002) 'Taking world cities literally', *City*, 6(3): 351–68.

Easthope, A. (2002) *Privileging Difference*. Basingstoke: Palgrave.

Ferguson, N. (2004) *Colossus: The Rise and Fall of the American Empire*. London: Allen Lane.

Fine, B. (2002) *What's Eating Actor-Network Theory?: the case for political economy in agro-food studies*, available at: http://www.soas.ac.uk/departments/departmentinfo.cfm?navid=495, accessed 15 September 2002.

Gibson-Graham, J.-K. (1996) *The End of Capitalism (As We Knew It): A Feminist Critique of Political Economy*. Oxford UK and Cambridge USA: Blackwell Publishers.

Hall, S. (1996) 'When was the post colonial?' in I. Chambers and L. Curti (eds) *The Post Colonial Question*, pp. 242–59. London: Routledge.

Hardt, M. and A. Negri (2001) *Empire*. Boston: Harvard University Press.

Harvey, D. (2003) *The New Imperialism*. Oxford: Oxford University Press.

Henwood, D. (2003) *After the New Economy*. New York: New Press.

Keating, M. (ed.) (2005) *Regions and Regionalism in Europe*. Cheltenham: Edward Elgar.

Layard, R. (2005) *Happiness: Lessons Form a New Science*. New York: Penguin Press.

Lovering, J. (1999) 'Theory led by policy: the inadequacies of the new regionalism', *International Journal of Urban and Regional Research*, 23: 379–95.

Markusen, A. (1999) 'Fuzzy concept, scant evidence, policy distance: the case for rigour and relevance in critical regional studies', *Regional Studies*, 33: 869–84.

Massey, D. (2004) 'Geographies of responsibility', *Geografiska Annaler*, 86(B): 5–17.

Peck, J. (2003) 'Fuzzy old world: a response to Markusen', *Regional Studies*, 37: 729–40.

Saad-Filho, A. and D. Johnston (2005) *Neoliberalism: A Critical Reader*. London: Pluto Press.

Sayer, A. (2000) *Realism and Social Science*. London: Sage.

Smith, N. (1996) *The New Urban Frontier: Gentrification and the Revanchist City*. London: Routledge.

Soja, E. (1989) *Post-Modern Geographies*. New York: Verso.

Wood, E. M. (2003) *The Empire of Capital*. New York: Verso.

Yeung, H. W. (2005) 'Rethinking relational economic geography', *Transactions of the Institute of British Geographers*, 30: 37–51.

19 Labour market geographies

Employment and non-employment

Anne Green

Introduction

This chapter provides a series of reflections on developments in labour market geographies, and associated concerns with employment and non-employment, over the last quarter century. It cannot and does not claim to provide a comprehensive overview, but rather it addresses selected key themes. While not restricted to the United Kingdom, it is there where the main emphasis lies – particularly with respect to policy.

At the outset a brief outline of the changing nature of labour markets over this period is provided to set the context for two main substantive sections of the chapter. The first main section deals with changing approaches and concerns of economic geographers. Five main issues are addressed under the 'approaches' sub-heading: (1) the change in emphasis from quantitative to qualitative methods; (2) the move away from empiricism towards theory; (3) the increasing weight placed on social and cultural issues; (4) the trend towards more detailed disaggregation; and (5) the role of geography and geographers in multi-disciplinary and inter-disciplinary studies of labour markets and labour market geographies. It is not possible to do justice within the constraints of space available to address the multiplicity of 'concerns' of economic geographers, so the main focus here is placed on the operation of local labour markets, with particular reference to four topics: (1) labour market adjustments; (2) the balance between migration and commuting; (3) the place of perceptions; and (4) the role of labour market intermediaries.

The second main section deals with changing policy issues. In the context of the opportunities for geographers offered by increased emphasis on evidence-based policy, amongst the issues highlighted are 'healthy' labour markets, and the shift in policy concerns from unemployment to non-employment, and from the 'quantity' to the 'quality' of employment. In turn, these issues relate to key policy questions, such as 'why' and 'how' concentrations of worklessness emerge, what should be the balance between supply-side and demand-side issues in labour market policy, and what policy levers are available at different geographical levels to influence outcomes. The final section of the chapter sets out some key features of the future agenda for researchers concerned with labour market geographies, and also a central question around which policy-relevant research could be focused.

Context

In the introductory chapter to 'Geographies of Labour Market Inequality', Martin and Morrison (2003) provide a useful overview of the changing world of work, which sets the context for academics and policy analysts concerned with the geography of labour markets. In simple schematic format, they identify four key forces of change: structural change (encompassing deindustrialisation, tertiarisation and privatisation); technological change (incorporating computerisation, informalisation and digitisation); globalisation (subsuming deepening, intensification and speed-up of international interactions and inter-dependencies); and deregulation and re-regulation (characterised by shifting power back to employers and a shift to active labour market policies). Amongst the main labour market impacts of these changes they identify the sectoral recomposition of employment, the skill recomposition of work, the gender recomposition of employment, union decline and new work relations, increased vulnerability to unemployment, casualisation and increased job insecurity, and widening of wage and income disparities.

This brief overview provides some insight into the reach and depth of the agenda, and the multiplicity of issues and impacts, facing economic geographers researching the geography of labour markets. It is not possible to touch on all of these issues here, but they are illustrative of the range of topics studied by geographers and other labour market analysts, and, more particularly, they set the context for key policy issues.

Changing approaches and concerns of economic geographers

Approaches

From quantitative to qualitative methods Over the last 20 years or so there has been some shift in emphasis amongst economic geographers (as amongst geographers generally) from application of quantitative to qualitative methods. This has occurred at a time when the availability of spatially-referenced data (from census, survey and administrative sources) of relevance to labour market studies has increased and the capacity and capability of computers and Geographical Information Systems (GIS) technologies to process and analyse such data has improved. Yet, arguably, knowledge of the existence, strengths and shortcomings of many data sources amongst many economic geographers remains underdeveloped. Quantitative, qualitative and, perhaps especially, mixed methods have an important role to play in geographical research, with the utilisation of qualitative alongside quantitative methods enabling rich insights to be gleaned into the operation of local labour markets. It is arguable that the pendulum perhaps swung too far in favour of qualitative methods, with concerns expressed by the Economic and Social Research Council (ESRC) about a lack of methodological rigour among the new generation of social scientists (quoted in Keylock and Dorling 2004).

From empiricism to theory Alongside the shift noted above, has been another important (and related) one from empiricism to theory. From the late 1980s, there has also been a shift in the relative importance of theoretical perspectives employed by geographers in local labour market studies (for a useful brief overview of theoretical perspectives see Martin and Morrison 2003), which in turn underlies the shift away from large-scale empirical studies towards more critical perspectives. The previously dominant competitive market perspective has been to some extent superseded by the imperfect market perspective, which emphasises heterogeneity by recognising the existence of a geographical mosaic of non-competing and segmented sub-markets and structural disequilibrium. Economic geographers have also increasingly adopted an institutionalist perspective emphasising local labour markets as socially embedded institutional spaces of formal and informal customs, norms and practices underpinning employment, work and wages, with some adopting a regulationist perspective (with the local labour market as a site of socio-political regulation). Hence, Peck (1989, 1996) called for attention to be focused on ways in which particular local intersections of labour supply, labour demand and state regulatory infrastructure are revealed in the form of different concrete outcomes in different places.

Increasing emphasis on social and cultural issues The shift towards enhanced emphasis on social and cultural issues evident in economic geography is part of a more pervasive trend in geography. As highlighted by Bauder (2001: 42), economic geographers have drawn on the concept of place to define not only the economic and political, but also the social circumstances that influence the spatial division of labour (as highlighted by Massey 1984 and Peck 1996). In turn, conceptualisations of place from social geography have also contributed to understandings of divisions in the labour market. The overall effect is to emphasise the social embeddedness of economic activity, the role of social relations in shaping economic interactions, the proactive role of place in labour market segmentation and the place contingent operation of local labour markets.

Towards more detailed disaggregation As highlighted by the imperfect market perspective, '*the* (local) labour market' suggests a unity absent. The geographical subdivision of labour markets has long been recognised (Goodman 1970), with, in any one local area, a multiplicity of sub-markets (characterised by varying degrees of permeability) demarcated by various criteria. Economic geographers have taken an increasing interest in how labour market geographies vary between different people in different places, and between different people in the same place. Conventionally, local labour market areas have been defined on the basis of aggregate commuting flows, but as more data and computing power have become available, there have been attempts to define such areas for different population sub-groups (Coombes et al. 1988). However, perhaps foremost in the trend towards more detailed disaggregation has been the increasing attention focused on gender as exemplified by the seminal work of Hanson and Pratt (1991, 1995), which explored how differences between men's and women's experiences of work are grounded, and constituted in and through space.

The role of geography and geographers Study of labour market geographies is not the unique preserve of geographers. Economists, sociologists and anthropologists, amongst others, have always had an interest in the role of space, and one senses a trend towards rediscovery of the importance of geography amongst other disciplines, at a time when, arguably, for geographers space has moved from centre stage to supporting role. Indeed, it is often not possible to identify easily the disciplinary background of those contributing to the analyses of local and regional labour markets. Often, the richest environments for research are at the boundaries of disciplines: hence, a vogue for multi- and inter-disciplinary studies of labour markets. For geographers such environments have proved tempting, and for some such forays have meant 'weighing anchor' from geography. As a result, it would seem that some 'geographers' are losing their identity – at least in the eyes of non-geographers with a traditional view of geographers being centrally concerned with space (and with maps – as exemplified by Dorling and Thomas 2004). Whether, if true, this matters for economic geography and geography more generally is another question.

Concerns

In order to illustrate some key developments in research concerning labour market geographies, four topics of relevance to the operation of local and regional labour markets are reviewed in this section. These topics have been selected to provide a flavour of the concerns of researchers, and to highlight new and emerging foci for study.

Labour market adjustments Adjustment processes are central to the operation of (local) labour markets. In line with the approaches and trends outlined above, at micro and local area scales economic geographers have emphasised the need for grounding analyses in a spatial perspective of labour market behaviour that recognises strong inter-connections of sub-markets through both geographical and occupational mobility, and the empirical significance of the specific ways in which adjustment processes operate (Gordon 2003). At local labour market area level the labour market accounts technique has proved insightful in highlighting how different areas have adjusted to job shortfalls. Following on from earlier applications (such as Owen et al. 1984), Beatty and Fothergill (1996) used the technique to examine the roles of migration, commuting, job creation and changes in labour force participation as mediating influences in labour market adjustment in British coalfield areas suffering job loss in mining. Of particular policy relevance here (as highlighted in the next section) was the significant reduction in labour market participation. The labour market accounts technique is avowedly quantitative and demanding of spatially disaggregated data, and as such it may be viewed as running counter to the trend towards greater emphasis on qualitative methods and on social and cultural perspectives. However, more qualitative case study approaches have proved valuable in emphasising the barriers to mobility and the rationale for immobility amongst sub-groups with relatively weak position in the labour market and strong social and cultural ties to specific places (Kitching 1990).

The balance between migration and commuting As highlighted above, spatial (im)mobility is a topic of central importance for the operation of local labour markets. Links between residences and workplaces have long been a central interest of economic geographers, population geographers and planners. At the intra-urban level one body of literature reflecting this interest is that on spatial mismatch, but the attention of geographers also extends to links between job access and labour market outcomes at inter-urban and inter-regional scales. Faced with a dearth of employment opportunities appropriate to his/her skills in a particular local labour market area a non-employed job searcher may remain non-employed, take a job for which he/she is over-qualified or search over a larger area and accept a more appropriate job at a greater distance. This latter option is likely to mean either extended commuting or migration. In a study of job access, workplace mobility and occupational achievement involving cross-sectional and longitudinal analyses, van Ham (2002) shows that urban structure influences labour market outcomes of individuals and urban structure itself is influenced by the spatial behaviour of workers. Good access to jobs leads to occupational achievement and reduces the need to be spatially flexible. A strategic residential location maximising the opportunity for commuting and so minimising the need for migration is particularly important for dual career households, combining two workplace locations with one place of residence. Such households, which have become quantitatively more important as the participation of women in the labour market and in high level jobs has increased, face complex location and mobility decisions, involving multiple work and non-work compromises and trade-offs (Green 1997; Hardill 2002). Changes in the organisation of work, enabling at least some workers to work away from the workplace for at least part of the time contribute to a more diverse choreography of working lives reflecting different configurations of, and responses to, lifestyle choices, work demands and other constraints.

The place of perceptions Until recently the role of area perceptions in shaping the behaviour of labour market and behaviour of individuals has been relatively neglected, despite the greater recognition of labour markets as social and institutional constructs. Yet historical patterns of socialisation and employment, residential location and segregation, and variations in spatial behaviour and local social capital contribute to different ways of 'knowing' the labour market by different people in different places. In a recent study of disadvantaged young people in Belfast involving testing knowledge of job concentrations in the city, the drawing of mental maps and focus group discussions, Green et al. (2005) show that limited mobility, geographical factors, religious factors and lack of confidence intertwine in complex ways to limit perceived opportunities and serve to create subjective opportunity structures that are a subset of all objective opportunity structures. Many young people restricted their options and chances of employment by discounting training and work opportunities in areas that were physically accessible but unfamiliar.

The role of labour market intermediaries A further relatively new topic of study for economic geographers is the role of labour market intermediaries. Labour market intermediaries broker the relationship between workers and employers through their involvement in three labour market functions: first, reducing transactions costs; second, building networks; and thirdly, managing risk. As such, they play an important role in shaping access to the labour market – both positively and negatively. In an overview using Silicon Valley as a case study, Brenner (2003) emphasises that labour market intermediaries are themselves varied, including temporary help firms, consultant brokerage firms, web-based job sites and professional employer organisations in the private sector; membership-based intermediaries such as professional associations, guilds and trade union initiatives; and public sector intermediaries encompassing institutions making up the workforce development 'system', education-based institutions providing adult education and customised job training for employers, and community organisations engaging in job training and placement activities. In the United Kingdom, as in the United States, labour market intermediaries are becoming more important in regional and local development and policy, and as such are a fruitful subject for further research.

Changing policy issues

Issues of employment, non-employment and the operation of labour markets at all geographical scales are of relevance to economic and social policy. In the United Kingdom labour market geographies are explicitly on the policy agenda, as exemplified by Public Service Agreement targets to reduce disparities in regional growth rates and to close the employment rate gap between the worst local areas and the national average. This suggests that there are opportunities for geographers to enhance their relevance by entering the policy debate and making their results accessible and informative to policymakers in the manner outlined by Markusen (2003).

The following sub-sections outline some of these opportunities in more detail and highlight important changes in emphasis on selected policy issues, which in turn offer new challenges to researchers with interests in labour market geographies.

Contributing to the evidence base for policy

Since the election of 'New Labour' in the United Kingdom in 1997 there has been increased emphasis on evidence-based policy. For researchers in academia and consultancies the raft of evaluation studies associated with New Deal programmes to combat disadvantage in the labour market and in other domains have offered opportunities to gain new insights into the ways local labour market operate in different places through investigation of what interventions work where, how and why; the role of labour market intermediaries; and the relative balance between supply-side and demand-side labour market barriers.

'Healthy' labour markets

'Healthy' has not conventionally been used as a descriptor in conjunction with 'labour market'. Yet with the publication of Frameworks for Regional Employment and Skills Action (FRESA) plans (resting on a robust and coherent evidence base) in the English regions in 2002 to serve as a focus for what needs to happen in a region to maintain and grow a 'healthy labour market', followed by the announcement in 2003 of the establishment of Regional Skills Partnerships to drive forward the regional skills agenda, 'healthy' labour markets comprise an important policy issue. There is no single accepted definition of what constitutes a 'healthy' labour market. Yet 'healthy' is a valuable term, and one that is challenging for researchers, in that it seeks to encompass something broader than economic concepts of 'efficiency', 'flexibility', 'rigidity' or 'tightness'. In essence, a 'healthy' labour market may be conceptualised as one that produces desirable results – both socially and economically, and which is sustainable over time. In a 'healthy' labour market there are skills and job opportunities at all levels, but the emphasis is on moving up the skills and value chain and ensuring that there is a progression route for those who choose to take it. Hence, a 'healthy' labour market has three key dimensions: (1) a strong demand side – in terms of quantity and quality of jobs; (2) a strong supply side – relating to the numbers and characteristics of people able to take those jobs; and (3) efficient and equitable functioning to bring together demand and supply. It also requires appropriate supporting conditions – including education, training, workforce development, benefits and welfare, housing and transport infrastructure. Creating a 'healthy' labour market depends on having a clear vision of the desired result, an understanding of the processes that might bring about that result, and how these processes might be influenced by public intervention. Once these are in place, indicators and targets for measuring the effectiveness of interventions and assessing progress may be set. Hence, there is ample scope for geographers and other labour market researchers to contribute to policy debates on growing and maintaining 'healthy' labour markets.

From unemployment to non-employment

In the days of 'slack' labour markets, job loss and unemployment were key concerns for researchers concerned with labour market geographies. Over time, however, unemployment has come to account for an ever-smaller proportion of non-employment. A range of different data sources have confirmed that unemployment has become an increasingly unreliable measure of labour reserve, and even more importantly from a geographical perspective, the 'more difficult' the local labour market, the smaller proportion of non-employment that unemployment captures, and the larger the share accounted for by long-term sickness (Mackay 1999). As conventional measures of unemployment have depressed the degree of spatial variation in labour reserve, so labour market analysts have been advocated the use of broader measures of non-employment (Green and Owen 1998) and the focus of policy attention has shifted increasingly to the inactive (who typically are more heterogeneous than the unemployed – since they include those looking after the home and family, who suffer sickness/disability, students, etc.).

From 'quantity' to 'quality' of employment

In contrast to 'quantitative' concerns about 'mass unemployment' in the mid 1980s, with a 'tightening' of labour markets there has been a shift towards greater policy emphasis on 'qualitative' aspects of employment, and on the role of skills as a key driver in regional competitiveness (the promotion of the 'healthy' labour market concept also reflects this trend). As a result the geography of occupations and skills has risen up the policy agenda. Geographers in private sector consultancy have been particularly influential and active here in entering and shaping the policy debate and making their results accessible and informative to policymakers. Local Futures' Regional Economic Architecture (Hepworth and Spencer 2003) adopts a four-fold knowledge-intensity classification to map and measure the demand- and supply-side of the regional geography of the knowledge economy in Britain, utilising an effective 'on one page' display and colour coding to indicate regional performance *vis-à-vis* the national average. This presentation is effective in highlighting key messages, such as the role of London as a knowledge economy hub and the importance of the public sector as a key driver of the knowledge economy in the North and Midlands. More detailed employment and labour force data at regional and sub-regional levels underpin these 'top' level results. Geographers in academia can and do contribute to the policymaking process, but perhaps not as much as they could (Martin 2001).

Spatial concentrations of worklessness

Despite the tightening of labour markets, spatial concentrations of worklessness at micro area level persist and remain a key concern for policy. So this is another topic where researchers concerned with labour market geographies have an opportunity to contribute to the debate in understanding how and why they emerge and are maintained. Many people living in such concentrations face multiple disadvantages, have low aspirations for work and study, and have extremely narrow travel horizons. Moreover, often there are two or three generations out of work in the same family in the same neighbourhood. Local area studies adopting both quantitative and qualitative methods, as well as evidence reviews and evaluation studies commissioned by government, have offered important insights into the processes underlying the lived experience of people in such neighbourhoods. A report by the Social Exclusion Unit (2004) has suggested that concentrations of worklessness happen for different reasons in different places, but that three main explanations apply: first, changes in the nature and location of jobs; second, the impact of housing market 'sorting'; and third, area effects (describing a situation in which once people live in an area with many people out of work their chances of finding work may be reduced simply by where they live). As highlighted by Gordon (2003) with respect to adjustment processes, understanding how spatial labour markets operate is crucial here. He argues that local concentrations of worklessness persist because they have become structural in character; such that they can only be removed by some combination of supply-side measures targeted at all the links in local processes that reproduce them, together with sustained full employment in the regions concerned.

Policy levers

In turn, this raises the question of what policy levers are available at different geographical levels to influence outcomes and which are appropriate in what circumstances. Again, there is scope here for economic geographers to contribute to the debate, through their understanding of labour market processes, governance and the role of intermediaries. In the United Kingdom there has been a trend towards greater devolution of responsibility for implementation and delivery of labour market related policies, so that regional and sub-regional partnerships are looking to identify and develop interventions at regional (and sub-regional) levels. Yet the importance of national policies and associated targets in shaping the activities of, and targets for, regional (and sub-regional) partners is clear, so placing limits on the scope and nature of regionally-specific interventions, unlike in countries with a federal structure and associated regional/state powers, such as the Germany and the USA.

Future agenda

Looking ahead, there is an ongoing role for economic geographers to monitor trends and changes in the labour market and their spatial impacts. Such monitoring inevitably has a substantial empirical component. Importantly, the emphasis here needs to be on 'flows' as well as on 'stocks'. To some this may seem mundane, but nevertheless such activity is valuable.

Many interesting and exciting avenues of investigation exist for researchers concerned with labour market geographies in the years ahead. Even if all such possibilities could be imagined, space constraints would preclude meaningful discussion. Hence, a single issue (and related sub-questions) is identified here.

'Mobility' and 'adjustment' have been emphasised in the preceding sections as concepts of fundamental importance in labour market geographies. One central question – of significance in theoretical, empirical and policy terms – around which research activity could be usefully focused, is:

What is the capacity for mobility and flexibility in labour markets?

'Mobility' and 'flexibility' are central to the adaptive capacity of regional and local economies. They are crucial components of a 'healthy' labour market with the capability for individuals and firms to move up the skills and value chain and adjust to change. Conversely, it is necessary also to understand the physical, technological and social constraints on mobility in the labour market that leave the weakest behind.

Issues of 'mobility' and 'flexibility' arguably take on renewed importance given two important and related challenges facing advanced economies (in Western Europe, in particular). The first is that posed by demographic change. In the light of low fertility and an ageing population, geographers and labour market analysts face important new questions relating to demography, labour supply and the role of immigration. The second is that posed by the forces identified by Friedman (2005) as 'flattening' the world at an unprecedented rate. So important related sub-questions include: What are, and what will be, the impacts of advanced technologies, outsourcing, offshoring, etc., for local, regional, national and international labour markets? And, Where is the dividing line between 'healthy' and 'unhealthy' mobility and flexibility?

References

Bauder, H. (2001) 'Culture in the labor market: segmentation theory and perspectives of place', *Progress in Human Geography*, 25: 37–42.

Beatty, C. and S. Fothergill (1996) 'Labour market adjustment in areas of chronic industrial decline: the case of the UK coalfields', *Regional Studies*, 30: 627–40.

Brenner, C. (2003) 'Labor flexibility and regional development: the role of labor market intermediaries', *Regional Studies*, 37: 621–33.

Coombes, M., A. E. Green and D. W. Owen (1988) 'Substantive issues in the definition of 'localities': evidence from sub-group local labour market areas in the West Midlands', *Regional Studies*, 22: 303–18.

Dorling, D. and B. Thomas (2004) *People and Places: A 2001 Census Atlas of the UK*. Bristol: Policy Press.

Friedman, T. (2005) *The World is Flat: A Brief History of the Globalized World in the Twenty-First Century*. New York: Allen Lane.

Goodman, J. F. B. (1970) 'The definition and analysis of local labour markets: some empirical problems', *British Journal of Industrial Relations*, 8: 179–96.

Gordon, I. (2003) 'Unemployment and spatial labour markets: strong adjustment and persistent concentration' in R. Martin and P. S. Morrison (eds) *Geographies of Labour Market Inequality*, pp. 55–82. London: Routledge.

Green, A. E. (1997) 'A question of compromise?: case study evidence on the location and mobility strategies of dual career households', *Regional Studies*, 31: 643–59.

Green, A. E. and D. W. Owen (1998) *Where are the Jobless?: Changing Unemployment and Non-Employment in Cities and Regions*. Bristol: Policy Press.

Green, A. E., I. Shuttleworth and S. Lavery (2005) 'Young people, job search and labour markets: the example of Belfast', *Urban Studies*, 42: 301–24.
Hanson, S. and G. Pratt (1991) 'Job search and the occupational segregation of women', *Annals of the Association of American Geographers*, 81: 229–53.
Hanson, S. and G. Pratt (1995) *Gender, Work and Space*. London: Routledge.
Hardill, I. (2002) *Gender, Migration and the Dual Career Household*. London: Routledge.
Hepworth, M. and G. Spencer (2003) *A Regional Perspective on the Knowledge Economy in Great Britain*, A Report for the Department of Trade and Industry. London: Local Futures Group.
Keylock, C. J. and D. Dorling (2004) 'What kind of methods for what kind of geography?, *Area*, 36: 358–66.
Kitching, R. (1990) 'Migration behaviour among the unemployed and low skilled' in J. Johnson. and J. Salt (eds) *Labour Migration*, pp.172–90. London: David Fulton.
Mackay, R. R. (1999) 'Work and nonwork: a more difficult labour market', *Environment and Planning A*, 31: 1919–34.
Markusen, A. (2003) 'Fuzzy concepts, scanty evidence, policy distance: the case for rigour and policy relevance in critical regional studies', *Regional Studies*, 37: 701–17.
Martin, R. L. (2001) 'Geography and public policy: the case of the missing agenda', *Progress in Human Geography*, 25: 189–210.
Martin, R. L. and P. S. Morrison (2003) 'Thinking about geographies of labour' in R. Martin and P. S. Morrison (eds) *Geographies of Labour Market Inequality*, pp. 3–20. London: Routledge.
Massey, D. (1984) *Spatial Divisions of Labour*. London: Macmillan.
Owen, D. W., A. E. Gillespie and M. G. Coombes (1984) 'Job shortfalls in British local labour market areas: a classification of labour supply and demand trends, 1971–1981', *Regional Studies* 18: 469–88.
Peck, J. (1989) 'Reconceptualizing the local labour market: space, segmentation and the state', *Progress in Human Geography*, 13: 42–61.
Peck, J. (1996) *Workplace: The Social Regulation of Labor Markets*. New York: Guilford Press.
Social Exclusion Unit (2004) *Jobs and Enterprise in Deprived Areas*. London: Social Exclusion Unit, ODPM.
van Ham, M. (2002) *Job Access, Workplace Mobility, and Occupational Achievement*. Delft: Eburon.

20 Technology, knowledge, and jobs

Edward J. Malecki

Introduction

Technology, knowledge and jobs are key ideas in the field of regional development. The unrelenting shift toward the service sector and toward high-technology and knowledge-oriented work is evidence that technology is changing. Many changes attributed to globalization are really outcomes of technological change – a result of enabling digital technologies, managerial techniques, and financial innovations. A recent innovation, the Internet, has loosened information asymmetries (if knowledge is power, many more are now powerful) and transformed shopping. Scholars, too, have benefited from the Internet and access to efficient communication and information. Google, the firm whose name we now use as a verb for searching the web, began only in 1998 (Google 2004).

I was fortunate to be among the (at that time) young 'industrial geographers' who attempted to remake economic geography during the late 1970s and early 1980s, integrating theory, largely from economics and management, with empirical analysis. Whether focusing on technology, as I and others did, or restructuring of the spatial division of labor, the goal we all sought was to understand better regional disparities in economic phenomena, such as employment, corporate facilities of different kinds, and entrepreneurial firms.

Many of these ideas were avant-garde in the late 1970s. More common at the time were studies of where companies – especially large companies – were located and how those locations changed over time, rather than of how companies themselves changed internally, with impacts on location, employment, and linkages. For myself and a few others, technology – or more precisely, technological change – was the principal answer. The places where firms did their research and development (R&D) were more likely to create and retain jobs than were places involved in routine manufacturing. This generalization continues largely to hold, in large part because the knowledge workers (as we call them today) involved in R&D have the intellectual, social, and financial wherewithal to innovate and assemble the resources needed to start new enterprises.

My contribution has been to understand how regional development is shaped by technology, including an array of indirect changes within firms and other organizations, in the networks that link organizations, and in the outcomes of these decisions on work and workers. My research has run the gamut from large secondary data sets to interviews with corporate executives, employees, and entrepreneurs, but with a common goal: to understand better how firms make decisions regarding their corporate facilities, how entrepreneurs make use of resources in both their local and non-local environments, and how small firms organize and manage inter-firm networks of suppliers, customers, and competitors. These questions can be answered only with data that are expensive and time-consuming to obtain.

My more recent work has been on the Internet and how firms and places use technology to manage and communicate across space. A web site is now a (not inexpensive) necessity for firms to attract customers, for places to attract tourists and investors, and more generally as proof of one's existence. If you can't be 'googled', you don't exist to growing numbers of people (Malecki 2002).

Influential geographers and non-geographers

Morgan Thomas and his students at the University of Washington (Geoffrey Hewings, Rodney Erickson, William Beyers, Gunter Krumme, Richard Le Heron, Guy Steed) were on the forefront of understanding regional development, integrating economics and geography. Their work was – and, in many cases, continues to be – central to the questions of regional economic development. Gordon Clark pioneered solid empirical research in economic geography on capital investment and finance. John Rees, using transatlantic connections, was the networker who convened several conferences that brought British and American economic geographers together, with the help of F. E. Ian Hamilton, Barry Moriarty, and Howard Stafford. Broader international networks were formed under the auspices of the various International Geographical Union (IGU) Commissions on industrial geography. Sadly, several of this group (Hamilton, Moriarty and Thomas) are no longer with us.

Peter Dicken (1986) summarized masterfully this work at the global scale, in a book that has been very successful in communicating geography to wider audiences through four editions (Dicken 2003). Dicken set a standard few others have matched in the empirical base he draws upon. It remains important for our research – and our theories – to match to a considerable degree what is out there in the real world.

Among non-geographers, economist Richard Nelson and the new evolutionary economics that he (with Sidney Winter) initiated in the 1970s are (explicitly or implicitly) essential to the understanding of technological change, institutions and economic growth (although apparently he never reads the work of geographers). Based more on what Nelson (1998) calls 'appreciative theory' than on the mathematical models of formal theory, the sub-field of evolutionary economics has much to offer to economic geographers. Those who call themselves evolutionary economists have done much to develop the set of concepts so important to understanding economic dynamics, whether at the national, regional, or local scale.

Theoretical advances

Research on jobs during the 1970s and 1980s revolved around the *spatial division of labor.* Doreen Massey's work was the lodestar, first expressed in Massey (1979) and then in book-length form in 1984 (Massey 1984, 1995). My own work in this vein used Dun and Bradstreet data to link corporate organization to locations of R&D and new firm formation, activities in the early stages of the product life cycle. I then turned to issues of entrepreneurship and of policy, summarized in Malecki (1997).

Others, including Thomas and his students, drew upon the economics literature to understand regional development and structural change, including shifts from manufacturing to services, and the greater utilization of technology and knowledge. A flurry of research on high-technology sectors ensued, which provided a better understanding of the role of research and development (R&D), of routine and nonroutine jobs, and of the corporate organization of multi-locational firms (Malecki 1991; Watts 1981). The shift to services as the leading economic sector in advanced economies remains difficult to study, since its outputs are frequently intangible and national data-gathering has lagged. Yet those outputs have become even more important and even more portable as digital files that can be bought and sold, pirated, and transferred effortlessly. Bill Beyers, Peter Daniels, and Peter Wood are among those who have led the way in research on services.

The language (or jargon) has changed a bit, but more fundamental aspects of regional development are timeless. The uneven destruction of jobs and the uneven creation of new ones will always be with us. The geographic scale has certainly changed: multinational or transnational corporations were only beginning in the 1970s to tap cheap labor pools in Asia and Latin America. Now, global production networks and outsourcing are much more common, yet we lack a full grasp of just which tasks, if any, might resist off-shore competition. In other words, are there any economic activities that must (or are most likely to) remain in advanced economies? Or are we doomed eventually to do little more than 'take in each other's laundry' and other personal services, while the most profitable and beneficial work is attracted to the sources of the cheapest labor? As an alternative, Pavitt (2003) provocatively suggests that systems integration, comprised largely of what are typically defined as services, is replacing manufacturing as the dominant activity in global industrial firms.

Measurement issues – what is important and how do we measure it – were and are critical. Employment, the most common measure of economic growth, is highly flawed in the context of jobless growth, or rising productivity without new jobs and in the realization that not all jobs are alike. Some are good jobs, with prospects for long-term advancement; others are dead-end jobs with low wages and little security in the face of race-to-the-bottom wage competition. Ann Markusen (1994) has pioneered our research methodologies in this area, with solid empirical work grounded in places.

A similar ambiguity holds for entrepreneurship and new firm formation: while many new firms such as Google exploit innovations and create new market opportunities, others are imitative and merely divide existing markets, especially the case with franchised retail outlets. Indeed, the distinction between entrepreneurship as a response to opportunity, as opposed to a response to necessity, is central to the benchmarking data collection effort of the Global Entrepreneurship Monitor (GEM).

Interesting empirical insights have emerged from the edges of geography, by people trained in other disciplines but who have embraced economic geography and contributed greatly to it, such as the late Bennett Harrison, Richard Florida, and Ann Markusen. They have been among the few researchers to connect economic process to real places and to write for wider audiences (Bluestone and Harrison 1982; Florida 2002, 2005; Harrison 1994).

Strengths and weaknesses of geographical analysis

What (many) geographers do best is to read widely and to grab an idea that originates elsewhere, and to enrich that idea with empirical detail. While economists and sociologists cite mainly within their own disciplines, geographers draw upon a wider cross-section of the social sciences and management studies. Geographers also tend to be more grounded in the real world than scholars in disciplines where an underlying theory or model has held sway for many years. Geography's own diversity truly reflects the diversity found in the world around us. Economic geography – or geographies – are products of research on the varieties of capitalism and of interactions of people and of institutions with those various systems (Lee and Wills 1997).

Geographers also try to grapple with unmeasured phenomena, such as linkages and interactions among firms, which range from aggregate intersectoral input–output flows to untraded interdependencies. Päivi Oinas and I have tried to map the numerous ways in which interfirm links affect the levels of knowledge and technology in various places (Oinas and Malecki 2002). These are difficult issues to study, relying on data that can be obtained only through interviews, compilation of anecdotal accounts, and detailed case studies; several such studies were included in Malecki and Oinas (1999).

Virtually lost from the early days is interaction with economics. In the 1970s, one could still find economists who were willing to think out loud outside the framework of neoclassical theory and its models. Such people are much more difficult to find today, even in regional science, where fewer cross-disciplinary conversations take place. Despite the emergence of a 'new economic geography', there seems to be much less interaction between economists and geographers than was the case in the 1970s.

Economists from outside the mainstream, in such sub-disciplines as evolutionary economics and industrial organization, occasionally take note of the work of economic geographers, as long as it appears in the right journals, such as *Research Policy* and *Industrial and Corporate Change*, in which few geographers have published. More geographical research should appear in those journals, rather than in outlets that focus on abstract, model-based research, whether labeled as the new economic geography or not. Only a few journals, such as *Regional Studies*, are cited by scholars from several disciplines.

Key research questions

The interesting questions concern the evolving and ever-changing interaction among labor skills, knowledge and technology, employment, and entrepreneurship. These processes are central to the competitiveness of places, and affect the range of choices regarding work for local populations.

Technology, broadly conceived as knowledge and its application, continues to be fundamental to regional change. It dramatically affects which jobs are created in which locations, and thus it profoundly influences the labor market choices of young people in specific locations. The pace of change and media attention to the migration of jobs to China and India only reinforces the importance of these processes. Our understanding of positive, cumulative development processes has progressed through various research 'fads', including high tech, regional innovation systems, and creative learning regions. These concepts have clarified our understanding of how regions work and how some places grow and develop while others do not.

In a setting of relentless competition and creative destruction, it is vital that we understand better the process of new firm formation – and of entrepreneurial success more generally. New jobs, new firms, and new industries occur most at the sites of innovation, whether they are based on R&D, new technology (such as IT), or on management innovations. Resources found in the (local and nonlocal) geographical environment are central to the variable chances that firms will be able to start, thrive, and grow.

Let me pose five key questions related to geography and entrepreneurship: (1) Is new firm formation equally likely everywhere? The *Global Entrepreneurship Monitor* (GEM) suggests that the answer is 'no' and that the reasons are related to cross-national variations in culture, policy, and many other variables. Technology-based firms and entrepreneurial universities also are part of the variation. (2) Why are some places entrepreneurial? We know a bit about successful places and innovative milieus and their wealth of role models, venture and seed capital, sources of ideas (universities, research), networks of interaction, cultures of entrepreneurship, and supporting institutions. But we have many more accounts of successful than of unsuccessful places. (3) Why are some places not so entrepreneurial? We have hints about unsupportive industrial structure and corporate organization, low levels of human capital, local cultures of distrust, and regulation, bureaucracy and corruption. (4) How can local environments be improved, and can governments help? Policies and politicians come and go, resulting frequently in policies too short-lived to be a basis for long-term development.

The fifth question re-opens the black box of technology in its new guise: How have the Internet and the rise of the creative class affected entrepreneurial environments? Although we know that delivering services is easier than delivering goods, the death of distance has not occurred. 'Handshakes' still require face-to-face interaction, even if 'conversations' do not (Storper and Venables 2004).

The creative class and its preferences for some places rather than others may appear to doom declining regions that cannot re-make and re-brand themselves.

The fact that not all jobs are equal – that some are unstable, have little upward mobility, do not stimulate one's creativity, or are otherwise unattractive in the long run – also needs continual research. Jobs are dumbed down, divided spatially, restructured, made redundant, replaced by computers, outsourced, and off-shored. Although we think now that knowledge jobs will withstand outsourcing, this might not be true.

Finally, knowledge – of all types – has grown to be essential to the character of good jobs and to the development of regions. Untraded interdependencies, which are essentially flows of tacit knowledge among firms, are among the most intriguing (Storper 1997). Data on tacit knowledge flows, like those on all untraded interdependencies and informal linkages, are impossible to track from any secondary data sets. Painstaking empirical work is needed simply to answer the important research questions.

Geographers and policy formulation

Geographers have much to contribute to policy, but few geographers are inclined to invest the time needed to build the level of credibility needed to influence policy. There are few of us compared to economists, and the latter are considered to know about economic matters, no matter how far removed from empirical reality their research actually is. Geographers, however, gain little from communicating only with each other. Publishing in other disciplines' journals and in more popular nonacademic outlets would raise awareness of what geographers know and do – and how it has some competitive advantage over the theoretical perspectives of the always-more-numerous economists.

The apparently anecdotal nature of much of the empirical research in economic geography also does not help to build credibility in policy circles. Case studies of a few, or interviews with a few dozen, can be easily dismissed as anecdotes, too few in number, unrepresentative, or unreliable. Perhaps meta-analyses of the many empirical studies would be useful.

The fact remains that geographers have spent far too little time contributing to public policy debates. This is not to say that geographers have not addressed public issues in their research. But that research is seldom made public beyond the restricted scope of scholarly journals, which few if any policymakers bother to read. The impact that GIS has had outside geography gives geographers of all stripes greater credibility than we have had in some time. The opportunity is here to take advantage of that heightened credibility to have an impact on the lives of people in the places we study.

References

Bluestone, B. and B. Harrison (1982) *The Deindustrialization of America: Plant Closings, Community Abandonment, and the Dismantling of Basic Industries.* New York: Basic Books.

Dicken, P. (1986) *Global Shift: Industrial Change in a Turbulent World*. London: Harper and Row.

Dicken, P. (2003) *Global Shift: Reshaping the Global Economic Map in the 21st Century*, 4th edn. New York: Guilford; London: Sage.

Florida, R. (2002) *The Rise of the Creative Class: And How it's Transforming Work, Leisure, Community, & Everyday Life*. New York: Basic Books.

Florida, R. (2005) *The Fight of the Creative Class: The New Global Competition for Talent*. New York: Harper Business.

Google, Inc. (2004) *Google History*, available at: http://www.google.com/corporate/history.html, accessed 13 October 2004.

Harrison, B. (1994) *Lean and Mean: The Changing Landscape of Corporate Power in the Age of Flexibility*. New York: Basic Books.

Lee, R. and J. Wills (eds) (1997) *Geographies of Economies*. London: Arnold.

Malecki, E. J. (1991) *Technology and Economic Development: The Dynamics of Local, Regional and National Change*. London: Longman.

Malecki, E. J. (1997) *Technology and Economic Development: The Dynamics of Local, Regional and National Competitiveness*, 2nd edn. London: Addison Wesley Longman.

Malecki, E. J. (2002) 'Hard and soft networks for urban competitiveness', *Urban Studies*, 39: 929–45.

Malecki, E. J. and P. Oinas (eds) (1999) *Making Connections: Technological Learning and Regional Economic Change*. Aldershot: Ashgate.

Markusen, A. (1994) 'Studying regions by studying firms', *Professional Geographer*, 46: 477–90.

Massey, D. (1979) 'In what sense a regional problem?', *Regional Studies*, 13: 233–43.

Massey, D. (1984) *Spatial Divisions of Labour*. London: Macmillan.

Massey, D. (1995) *Spatial Divisions of Labour*, 2nd edn. London: Macmillan.

Nelson, R. R. (1998) 'The agenda for growth theory: a different point of view', *Cambridge Journal of Economics*, 22: 497–520.

Oinas, P. and E. J. Malecki (2002) 'The evolution of technologies in time and space: from national and regional to spatial innovation systems', *International Regional Science Review*, 25: 102–31.

Pavitt, K. (2003) 'What are advances in knowledge doing to the large industrial firm in the "new economy"?', in J.F. Christensen and P. Maskell (eds) *The Industrial Dynamics of the New Digital Economy*, pp. 103–20. Cheltenham: Edward Elgar.

Storper, M. (1997) *The Regional World*. New York: Guilford.

Storper, M. and A. J. Venables (2004) 'Buzz: face-to-face contact and the urban economy', *Journal of Economic Geography*, 4: 351–70.

Watts, H. D. (1981) *The Branch Plant Economy: A Study of External Control*. London: Longman.

Index